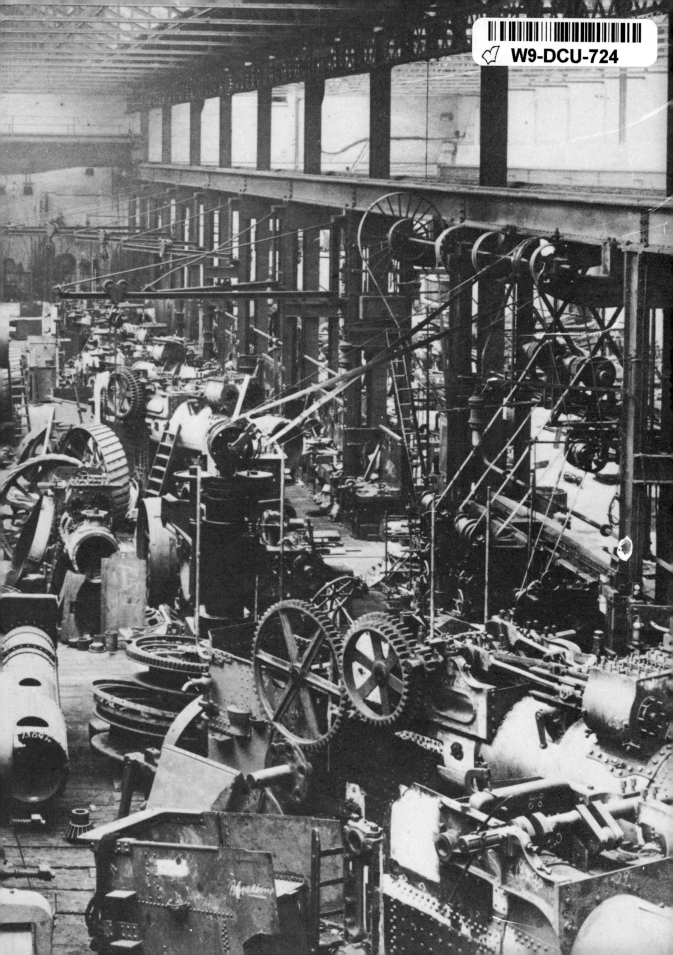

THE POWER OF
STEAM

AN ILLUSTRATED HISTORY OF
THE WORLD'S STEAM AGE

ASA BRIGGS

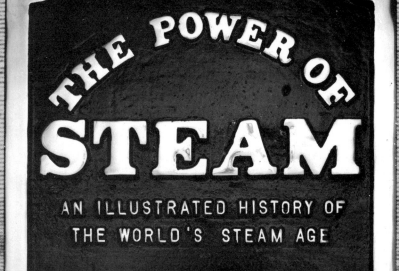

THE POWER OF
STEAM

AN ILLUSTRATED HISTORY OF
THE WORLD'S STEAM AGE

ASA BRIGGS

THE UNIVERSITY OF
CHICAGO PRESS

The University of Chicago Press,
Chicago 60637
Michael Joseph Ltd., London WC1
© 1982 by Bison Books Ltd.
All rights reserved. Published 1982
Printed in Hong Kong

89 88 87 86 85 84 83 82   1 2 3 4 5

Conceived and prepared by
Sheldrake Press Ltd.

EDITOR: Simon Rigge
Picture Editor: Karin Hills
Designer: Roy Williams
Consultant: K. M. Brown
Picture Researchers: Dee Robinson,
Jim Latter
Editorial Assistants: Annabel Lloyd,
Susan Marshall

**Library of Congress Cataloging in
Publication Data**

Briggs, Asa, 1921–
   The power of steam.
   Includes index.
   1. Steam engineering—History. I. Title.
TJ275.B825 1982   621.1'09   82-40321
ISBN 0-226-07495-1   AACR2
ISBN 0-226-07497-8 (pbk.)

Front endpaper: Partly assembled traction engines and steamrollers fill the erecting shed at the works of John Fowler & Co., Leeds, in 1912.

Back endpaper: In an engraving of the 1840s, the Lockwood Viaduct of the Huddersfield and Sheffield Railway, complete with a train crossing, provides a noble addition to a peaceful landscape.

Previous page: 'The Industrial Aspect of Leeds from Richmond Hill,' engraved in 1885, offers a classic expression of the age of steam.

This page: A giant steam tractor ploughs in California in 1904. American-built by Benjamin Holt, it was one of the earliest steam engines to use caterpillar tracks, a British invention.

CONTENTS

THE CONQUEST OF THE MATERIAL WORLD

The *Rocket* soars into the skies;
With all our hopes and fears she flies
With beef and coal, with beer and blocks
Of ordinary shares and stocks;
We set no limit to our dream
Till all the vales shall pant with steam,
And puffs of smoke on hill-sides proud
Shall wander lonely as a cloud.

E. V. KNOX, *The Steam Givers*

In a sketch entitled 'Concurrence', an artist of the 1870s compares the two worlds of steam and animal power. Since ancient times the maximum speed a man could average over a considerable land journey had hovered at between one and five miles an hour. Now average speeds of more than fifty miles an hour were not uncommon on the railways.

During the last four hundred years,' wrote the American historian John U. Nef in 1941, in an important article on the rise of industrialism, 'the Western peoples have concerned themselves, to a greater degree than any other peoples before them, with the conquest of the material world.' In the story of that conquest the introduction of new forms of power was of crucial importance, and the greatest breakthrough of all was the first: the discovery and application of the power of steam.

For centuries society depended on human and animal muscle, wind and water, and all such natural forms of power are still in use. It was not until the beginning of the 18th century, when the first steam engines were built, that men began to break free from this limiting dependence – a process that originated in England and spread by stages to the rest of the Western world and beyond. In the process the Western peoples became increasingly dependent on mineral resources, firstly coal, and on technical and scientific advance.

By 1941, the year of the entry of the United States into the Second World War, when Nef wrote, advanced countries both in peace and in war relied for their very existence on oil and electricity, and by then, too, highly sophisticated scientific work was already in progress on the preparation of an atomic bomb. The dropping of the first atomic bombs

Climbing up the rim of a large windlass and heaving at its spokes, a dozen men struggle to raise a heavy stone at a quarry near Gentilly in France. The principles of muscle- and water-powered machinery had been known as long ago as 250 B.C., and continued to be exploited long after this scene was painted in 1826.

on Japan in 1945 was a supremely dramatic demonstration that new forms of power could destroy as well as create. In more recent years, since the Arab-Israeli war of 1973, we have been aware not only of the darkening presence of nuclear weapons, but of an oil crisis which has already influenced economics and politics, and which threatens what have come to be accepted as normal ways of life.

It is within such perspectives that this book, *The Power of Steam*, has been conceived. A book called 'Steam Power' – and there have been many such since the 18th century – would find its place in a library with books on mechanical engineering. It would describe how steam engines and steam turbines work, how their efficiency can be measured, and how their construction, use and efficiency are related to thermodynamics. A book called 'The Power of Steam', however, can have different connotations, and this one has. It deals with the inventive force that lay behind the discovery and exploitation of steam power, with the hopes and fears that accompanied its development, and with the challenge of finding and applying alternative forms of power.

'Every step forward in human history has been accomplished by laments that it could not be made,' the British historian E. L. Woodward pointed out in 1945, commenting on the effects of the atomic bomb, 'and that, if made, it would have results leading perhaps to catastrophe.'

In a woodcut published by Georgius Agricola of Saxony in 1556 (below), five men operate a haulage machine powered by a reversible water-wheel. With skill, two leather buckets could be filled alternately from a well. By the 1880s, one man operating a conical-drum steam winder like that on the right could, with two helpers to load and unload, raise coal tubs from deep mines at speeds of at least sixty feet a second, with intervals of less than sixty seconds between winds.

The steam engine was no exception. There had to be persistent effort before it was perfected and there was much dispute about its likely effects – whether it would liberate or enslave; whether it would destroy the environment; whether through locomotion it would pull the world together or ruin its peace.

From the late 17th century, one point was clear to all far-sighted contemporaries: the development of the engine would have universal, and not merely local or national, implications. The process of invention was itself international, as the process of discovery and development in atomic energy was also to be. That is why this book, unlike most books on steam, is not restricted to a study of one particular part of the world. It seeks to relate the story of the power of steam – in its different phases – to the unfolding of world history.

The age of the steam engine was relatively short. It began, however, before the seminal work of James Watt, who figures most prominently in all accounts of its achievements – so prominently indeed that it is difficult to consider his particular achievement in perspective. Chapter 1 of this book deals with people who are sometimes called too simply 'Watt's predecessors'. These men were either scientific explorers interested in experiments with steam, or more practical men seeking to meet one of the major challenges of their time: pumping water out of flooded coal or metal mines. The scientific explorers were to be found in many countries; the more practical men, Thomas Savery and Thomas Newcomen outstanding among them, were in the first instance English.

Chapter 2 on Watt himself deals with a wider challenge than that of the mine: the challenge of harnessing power to every kind of machine. The progress of mechanization (a new word) in the 18th century generated a demand for power on an unprecedented scale. In the textile industry in particular, where new machinery was introduced in both spinning and weaving, there was a restless struggle by the 1770s to obtain 'power and more power'. The challenge of the mill became the starting point of what soon began to be thought of as the 'industrial revolution', a revolution which inevitably transformed other industries besides textiles. 'Industry', regarded increasingly as a major sector of the economy, not as it had been earlier in history as a quality of human beings, turned increasingly to steam during the 19th century. So for a time did the most advanced agriculture. And as the number of steam engines multiplied, the demand for iron and coal multiplied too.

It was steam locomotion, however, which was, and still is, usually taken to be the chief of the 'triumphs of steam'. For Henry Frith, who wrote a book with that title at the beginning of the 20th century, 'the iron horse' was the 'king of beasts; he drives the lion and the elephant from his path, and disturbs the polar bears in their icy fortresses'. Much of this kind of rhetorical language is quoted in Chapter 3 on 'the gospel of steam'. It should be set alongside the hard but often misleading statistics collected officially and unofficially. In this book, and particularly in this chapter, I have drawn heavily on the often neglected writings of contemporaries, as I did in my books *Victorian Cities* and *Ironbridge to Crystal Palace*. Steam had the power to move men as well as machines, to inspire prophets as well as businessmen, and to generate controversy as well as to establish new routines of life and work. The reactions of contemporaries to the new technology, part of what may be

Steam-powered cotton mills tower above the River Irwell, belching smoke into the Manchester sky, in an evening scene of about 1830. Cotton textiles were the first product to be made in factories employing hundreds of people. Even in 1851, when the advertisement (right) appeared, most industrial workers in England were employed in firms with less than fifty employees.

called the culture of steam, are best expressed in their own words.

The chapter on the gospel of steam precedes Chapter 4 on railways and steamships, for although the arrival of steam transport stimulated much enthusiastic talk about the 'annihilation of distance', thoughtful people had foreseen its coming and foretold its benefits decades before. Already by the year 1830, when George Stephenson's *Rocket* made its pioneer journey along the famous new Liverpool to Manchester Railway, commentators were selecting other more far-reaching themes for attention. One writer captured the optimism of the time when he described the steam engine as 'the latest born but the greatest of the sons of Vesta' and called it 'the fourth estate of the realm'. 'It is the real and effective balance in the state,' he went on; 'it maintains the credit of the national debt, it is the thunderbolt of war, and the fruitful olive of peace.'

Similar approbation was expressed in other countries. Thus, the great French critic Ernest Renan asked in 1848 what speculative discovery of the mind had exerted as much influence as the discovery of steam power, and the French Academy three years earlier had set steam power as the subject for its annual poetry competition. The winner and the runner-up both identified steam with progress.

The gospel of steam did not, however, win universal acclaim. Another French writer, Auguste Barbier, lesser known than Ernest Renan but equally eloquent, chose a cotton mill as a scene from the inferno and turned words like bobbin and pistons into tokens of horror. In both European and American literature, railways, far from being emblems of progress, could be seen as symbols of death. Although the material benefits they brought were usually recognized also.

In politics and economics – political economy was the contemporary word which linked them together – a clear division appeared between champions and opponents of steam. It was out of experience of early industrialization based on steam that manufacturers came to demand free competition and free trade, and workers to press for co-operation and factory legislation. Of course, technology, old or new, can never be separated from its economic and social context, and the context of the diffusion of steam power – what the American writer Lewis Mumford called 'carboniferous capitalism' – was very different from that in which atomic power or even electricity developed. Manufacturers, often new, uncultivated men, owned the machines – their capital – and made profits; workers, who had seldom been employed for regular days of work in factories before, laboured and received wages. This division of interest influenced attitudes towards the new technology as much as the inherent characteristics of the technology itself.

The rhythms of domestic production had been intermittent, while those of the factory were regular and persistent. Women and children, the main element in the labour force of the cotton textiles industry, the first to be taken over by steam, got tired as they worked their long hours; the steam engines never tired.

From the employers' point of view, it was economically so important to keep power-driven machinery in regular operation that, not surprisingly, fines were imposed on workers who jeopardized the routine – by arriving late or going off early or by opening windows or leaving the room. These were first-generation problems, but later in the history of steam power in factories employers faced further problems in raising machinery speeds as the amount of power increased. Andrew Ure, who looked at developments not from the workers' but from the employers' point of view, defended the newly established discipline and welcomed the speeding up of work. Human beings, he declared, were trained to 'renounce their desultory habits of work'.

The introduction of steam power marked an obvious divide between pre-industrial and industrial society and factory discipline was one of its key innovations. Karl Marx, who criticized Ure, himself agreed that if the dynamic of private enterprise had not been at work in the 18th century, it would have been impossible to introduce such an innovative new factory technology. He noticed, too, that the far-reaching implications of continuing technological improvement included some which had been foreseen, such as the growth in the production of material goods, and some which had been unforeseen, particularly the eventual emergence not only of a new labour force but of a working class.

As an advertisement, the engineers John and Henry Gwynne of London included in their 1876 catalogue this scene showing the application of their 'Patent Portable Direct-acting Pumping Engine for Irrigating Chinese Paddy Fields'. British manufacturers led the world in the export of steam-powered machinery.

Chapter 5 describes and discusses the growth of steam power and its underlying technology in different parts of the world and assesses its economic and social consequences. The process was accomplished over decades and with different chronologies not only in different countries but in different parts of the same country. There were some countries – notably the United States and France – where water power remained more important than steam power in the early decades of industrialization. As Nathan Rosenberg has emphasised in his fascinating *Perspectives on Technology* (1976), as late as 1869 steam power accounted for barely half of primary power capacity in United States industry, with water providing the other half. In other countries such as Japan, rapid industrialization began only after the age of steam had given way to the age of electricity.

In spite of wide divergences in national chronologies, the impact of steam was universally felt. On both sides of the Atlantic, metaphors derived from steam are still common not only in the language of writers but in everyday speech. We frequently speak of 'getting up steam', 'working off steam', 'running out of steam' or of 'full steam ahead'. The Oxford Dictionary gives 1826 as the first year when 'steam' was used figuratively to imply 'energy', 'go', driving power and the like, and this sense of steam has been carried into the non-English speaking world. For example, a Japanese-English dictionary gives half a dozen steam metaphors which made their way across the Pacific sixty or seventy years later.

By then, however, there were alternative new power technologies, notably electricity and the internal combustion engine. As the alternative technologies developed, enthusiasm and nostalgia for steam grew – the theme of Chapter 6, the last in the book. It is a feature of technologies that they captivate most when they are disappearing or dead. Stagecoaches, sailing ships, waterwheels – all had their days of romance in the 19th century. In the 20th, steam engines, particularly locomotives, have come to appear as things of beauty to large numbers of people, including many who did not live through the age of steam. Thousands visit steam rallies or work steam engines and locomotives in their leisure time, and almost as many books have appeared on vintage steam as were published on steam power in its beginnings and in its heyday.

In spite of all the current interest in steam, few recent books on steam encompass both the economics and the aesthetics of the subject. There is a great gulf between albums of photographs of locomotives and steam engines, some of them beautifully produced, and such severe quantitative studies by economic historians as G. N. von Tunzelmann's *Steam Power and British Industrialisation to 1860* (1978), which analyses the economics of technical change. My own book seeks to bridge the gulf.

Two branches of the subject have already been very well covered. All historians of steam power owe an immense debt to Henry W. Dickinson who, followed by L. T. C. Rolt, gave new life to the history of technology. Dickinson's *A Short History of the Steam Engine* (1938), one of several which he wrote on such themes, is a key book. So, too, is Rolt's *Thomas Newcomen: the Pre-History of the Steam Engine* (1963), a book which is as careful to do justice to Newcomen, an outstanding pioneer, as Dickinson was to do justice to Watt in the equally remarkable study he produced with Rhys Jenkins, *James Watt and the Steam Engine* (1927). The technical history of the steam engine, including the compound engine, needs little further elaboration, although there is still much practical work to carry out in the field of industrial archaeology.

The other branch which has borne particularly good fruit is the study of the place of science in technology. This subject has been well tended by A. E. Musson (who wrote an important revisionist introduction to a new edition of Dickinson's *Short History* in 1963) and Eric Robinson (who, with Douglas McKie, edited Watt's letters to the great 18th-century chemist, Joseph Black). Together Musson and Robinson selected from Watt's correspondence and notes a set of essential documents on 'the steam revolution', which they published in 1969.

The result of their work has been a re-evaluation of the role of science in the introduction of steam power. For many 19th-century writers, the steam engine was the product of 'untutored genius', owing more to mechanics than to men of science: 'there is no machine or mechanism,' wrote John Stuart in his much quoted *History of the Steam Engine* (1824), 'to which the little that theorists have done is more useless. The honour of bringing it up to the present state of perfection, therefore, belongs to a different and more useful class.' This view was echoed in 1968, the year Musson and Robinson challenged it, by Maurice Daumas in the introduction to Volume III of the important *Histoire Générale des Techniques* which he edited: 'the steam engine,' he wrote, 'like the internal combustion engine a century later, was born without the assistance of science.' Musson and Robinson have forced the historian to reconsider such verdicts, drawing attention to the pace-making experiments of 17th- and 18th-century science. Yet practical men produced the first steam engines, and both invention and development preceded the discovery and elaboration of the relevant scientific theory of the engine itself, thermodynamics. This theory was worked out by different men in France, Germany and England in the 1820s and 1840s. The question of just what their relative contributions were has been fully discussed by Yehuda Elkana in *The Discovery of the Conservation of Energy* (1974), a careful and critical account of the meeting of converging lines of thought, which takes as its motto an older verdict: 'Science owes more to the steam engine than the steam engine to science.'

Andrew Ure in his influential but, as we have seen, one-sided book

In an engraving of 1831, two pumping engines are at work draining the Dolcoath copper mine in Cornwall, but the ore is still crushed and raised in the shaft by hand. Thomas Newcomen, inventor of the first successful piston-in-cylinder steam engine, and Richard Trevithick, who built the first locomotive, were both engineers from the West Country.

Philosophy of Manufactures (1834), was frank enough to state that 'the university man, preoccupied with the theoretical learning, is too apt to undervalue the science of the factory'. I do not myself feel in danger of such undervaluation. I was brought up in an industrial environment transformed by steam, and in the town where I was born, Keighley in the West Riding of Yorkshire, 18th-century inventors contributed to the development of the power loom and 19th-century workers protested against the tyranny of steam.

Even academic life has its connections with technical history. In the library of Worcester College, Oxford, where I am now Provost, there can be found the oldest drawing of a Newcomen engine, re-discovered in 1925. It bears the inscription 'The Engine for Raising Water (with a power made) by Fire' and was drawn in 1717 by Henry Beighton, a Fellow of the Royal Society, who knew as much about the application of the Newcomen engine to the pumping of mines as Ure did of the application of the Watt engine to driving textile factories.

There will be few theoretical formulae in this book. I shall be concerned more with practical applications and imaginative responses, recalling Henry Dunckley's words in his prizewinning essay *The Charter of the Nations* (1854) in which he described the steam engine as 'an invention which seems the necessary complement of every other, which renders art prolific, and arms the softest touch of genius with demiurgic power'.

The facts about steam power are striking in themselves, and were often tabulated; it is not an accident that the growth of steam power and the growth of statistics coincided in time, for statistics offered the most convenient indicators of the conquest of the material world. The folklore is equally striking, leading us back to lost ways of thinking and feeling. This book is an exercise both in recovery and in interpretation, spanning an even longer period than the four hundred years identified by Nef and ending with the fate of steam in my own lifetime.

ASA BRIGGS

Worcester College, Oxford

At an early motor show in Tunbridge Wells in 1895, a crowd gathers round a steam horse attached to a carriage. Designed by Count de Dion and M. Bouton of France, who also exhibited one of their petrol-engined cars, this vehicle ran on coke fed in through a stoking chute visible on the bonnet. The gases from the fire passed down an exhaust duct that came out at the back.

In the last days of steam on the American railroads, two 4–8–4 Reading Line locomotives (right) thunder with a heavy train up the 1 in 80 gradient to Locust Gap, between Shamokin and Tamaqa, Pennsylvania. Steam, by conquering the vast continental distances of the New World, opened up the moving frontier of the United States.

CHAPTER ONE

MAKING STEAM WORK

Great inventions are never, and great discoveries are seldom, the work of any one mind. Every great invention is really either an aggregation of minor inventions, or the final step of a progression. It is not a creation but a *growth* – as truly so as that of the trees in the forest. Hence, the same invention is frequently brought out in several countries, and by several individuals, simultaneously. Frequently an important invention is made before the world is ready to receive it, and the unhappy inventor is taught, by his failure, that it is as unfortunate to be in advance of his age as to be behind it.

Inventions only become successful when they are not only needed, but when mankind is so advanced in intelligence as to appreciate and to express the necessity for them, and to at once make use of them.

ROBERT H. THURSTON,
A History of the Growth of the Steam Engine (1883)

A Newcomen engine stands idle, along with two bystanders, at Nibland Colliery, Nottinghamshire, in 1855. Thomas Newcomen's first recorded engine, erected to pump out a Staffordshire colliery in 1712, marked the beginning of the steam age. Many engines of his design were later adapted, like this one, to perform winding duties, and a few long-lived relics were still working in the early 20th century.

The discovery and exploitation of steam power began long before the four hundred years of conquest of the material world, perhaps the main theme in the history of modern times. It was in the ancient world that the first experiments with steam gadgets were made, not to increase the production of material things, but to achieve 'surprising results'. Magic and religion provided the context, not industry and political economy; and ingenuity was prized more than utility.

Almost every historian of the steam engine begins with Hero of Alexandria, who described in his *Spiritalia seu Pneumatica* (c. 60 AD) how 'vapours of water' could be harnessed to create contrived effects. Jets of steam could make a horn blow or a bird sing or a hollow ball whirl around, and in religious rituals an extra element of mystery could be achieved through the ingenious use of cylinders, levers, valves, stop-cocks and pistons. Heat applied to a liquid could make a libation pour on an altar; applied to air it could make temple doors open and close. The air, heated by an altar fire, drove water out of a sealed container into a bucket which descended on a cord and opened the doors. When the air cooled, it contracted and sucked water back out of the bucket, so closing the doors.

There are four manuscript copies in the British Museum of Hero's *Pneumatica* written in the 15th and 16th centuries, when there was growing interest in Faustian dreams of power. A 19th-century translation of Hero was published also in London in 1851, the year of the Great Exhibition in the Crystal Palace, when steam power had become a reality. It was only in the light of what came later that historians of technology came to see Hero's whirling ball, or aeolipyle, as a simple form of reaction turbine.

Nineteenth-century historians of technology tended to move straight from the ancient world to the age of 'the renaissance' and to leave out the middle ages altogether. H. W. Dickinson also omitted them from his *Short History of the Steam Engine*, the best 20th-century account. Yet as Lynn White and Bertrand Gille have shown in studies of mediaeval technology, 'the middle ages introduced machinery into Europe on a scale no civilisation had previously known', and if we wish to fit the evolution of the machine into the story of economic and social development – and the demand for power which it generated – we cannot leave the middle ages out. A leading historian of that period, E. M. Carus Wilson, did not hesitate to apply the term 'industrial revolution' to the mechanization in mediaeval England of fulling (the process of cleaning and thickening cloth) through the development of water power. The full exploitation of water power, the force of which had been recognised in the ancient world, was a necessary prelude to the development of steam power and the industrial revolution as we know it.

Karl Marx got the order right in Chapter XV of his *Capital*, 'Machinery and Modern Industry', in which he wrote that the starting point of modern industrialization was the machine which 'supersedes the workman who handles a single tool, by a mechanism operating with a number of similar tools, and set in motion by a single motive power, whatever the form of that power may be'. He went on to note that 'the steam engine itself, such as it was at its invention, during the manufacturing period at the close of the seventeenth century, and such as it

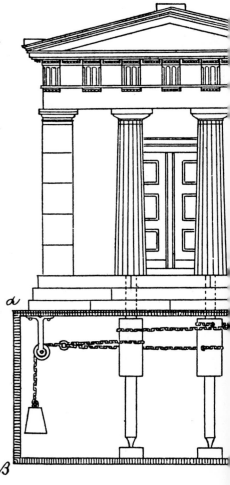

The secret of the automatic temple doors described in the 1st century A.D. by the Alexandrian writer Hero is revealed in this diagram from a 19th-century edition of his Opera. *When a sacrificial fire is lighted, air in the hollow altar expands and passes through a pipe into a sealed sphere below. It drives water from the sphere into a bucket which descends on a cord, turning two drums, raising a counterbalance and opening the doors; when the fire is put out, the air contracts, sucking water back out of the bucket and allowing the counter-balance to fall, so closing the doors.*

continued to be down to 1780, did not give rise to any industrial revolution. It was, on the contrary, the invention of machines that made a revolution in the form of steam engines necessary'.

Water power drove the fulling revolution into the countryside where streams and rivers flowed. If in retrospect it has seemed to carry as much romance, or even more, than steam power, its introduction was just as capable of generating conflict. Some of the complaints which were to be heard in the 18th century about the social effects of steam-powered machinery had been heard in the 13th century too. Water power brought opportunity and prosperity to the country as a whole, but in certain old sectors of industry it produced discontent. So, too, did the introduction of windmills. Yet the widely held idea that the men of the middle ages feared to tamper with nature's forces was not universally evident. 'Not all the arts have been found,' a Dominican friar pointed out in Pisa in 1306, 'we shall never see an end of finding them.'

In the 16th and 17th centuries a spirit of observation and enquiry revived old interests in steam and generated new ones. In 1606 an Italian 'natural philosopher', or scientist, Giovanni Battista della Porta of Naples, described an experimental water pump similar in principle to Hero's temple-door-opening device, in which steam rather than air forced the water out of a closed container. He also reported how a flask full of steam, when upturned in a bowl of cold water, would draw up water. (As the steam cooled and condensed, it created a vacuum which the water rose to fill.) At about the same time, Salomon de Caus, a French landscape gardener, who lived at various times in England, Italy and Germany and installed waterwheels, pumps and other devices to work fountains in the gardens of wealthy clients, built a simple steam-pressure apparatus by which he demonstrated that 'water will mount by the help of fire higher than its own level'. The apparatus was a hollow ball similar to Hero's aeolipyle, except that instead of turning round and round, it spouted a fountain of water.

With similar ingenuity, an Italian, Giovanni Branca, conceived of a machine for propelling a wheel by a blast of steam and described it in a book published in Rome in 1629. The Reverend Dionysius Lardner, one of the early 19th-century writers on steam, noted in 1827 that it had 'no analogy whatever to any part of the modern steam engines in any of their various forms', although a hundred years later H. W. Dickinson was able to call it an early impulse turbine.

De Caus never discovered any practical use for his steam fountain, but he nonetheless stands out as the first in a line of engineers concerned with the pumping of water, and was fittingly commemorated by a French poet who in 1842 wrote verses entitled 'Salomon de Caus or the Discovery of Steam'. It was the challenge of pumping water out of deep mines – in an age when mines were becoming economically important – that turned the search for steam power from a diversion into a necessity. The sense of necessity was stronger in England than in Italy and France, for early industrial growth in England – described by Nef as another era of 'industrial revolution' – already depended not just on wind and water, but on coal and metals.

The early 17th century was a time, too, when there were experiments with steam in laboratories and dreams of designing machines 'to propel vessels without oars or sails'. As the historian of science A. C. Crombie has put it, the expansion of trade and development of mining encouraged philosophers to develop 'a vigorous interest in the study of the technical process of manufacture, and this helped to unite the mind of the philosopher with the manual skill of the craftsman'.

In 1606 a contemporary of de Caus, James I's great public servant Francis Bacon, proclaimed boldly that 'the empire of men over things depends wholly on the arts and sciences' and that 'the true and lawful good of the sciences is none other than this: that human life be endowed with new discoveries and powers'. In his description in his *New Atlantis* (1620) of Solomon's House – a kind of laboratory set aside for the discovery of 'the knowledge of causes and secret motions of things' – he found room for 'engines' to be 'prepared for all sorts of motions'; and the Royal Society, founded in 1662, provided if not a Solomon's house, at least a useful forum where men of science could meet. It had a practical emphasis too. Its first historian, Bishop Sprat, described how it proceeded by 'action rather than discourse', adding that 'the genious [*sic*] of experimenting is so much dispersed that all places and corners are busy and warm about the work'.

Bacon's recognition of the need for a union of the 'arts and sciences' was obviously shared by many other people in the 17th and 18th centuries, not all of them or even most of them immediately concerned with the practical problems of mining. Royal or aristocratic patrons made demands of their own. In 1641 Cosimo de Medici, Grand Duke of Tuscany, required his engineers to make a suction pump to draw water from a depth of fifty feet. When the engineers found that the water would rise no more than about twenty-five feet up their pipe, leaving an unfilled vacuum above, they appealed for an explanation to Galileo. A lame reply from the great man led others to investigate further.

It was clearly not enough to argue, as philosophers had been accustomed, that nature abhorred a vacuum and that wherever a vacuum appeared air or water would rush to fill it. A vacuum existed and water was not rushing to fill it. Out of this scientific impasse came the discovery that air has weight and can exert pressure. Water rose to fill the vacuum in a pipe because it was pushed, but it would go only as high as the level where it balanced with the air, namely twenty-five feet. The demonstration of 'the spring and weight of the air' was made in Italy by Evangelista Torricelli and Vincenzio Viviani, both followers of Galileo; significantly it was made, too, in France by the philosopher Blaise Pascal and independently in Germany by Otto von Guericke, mayor of Magdeburg. This was a characteristic example of how great discoveries tend to be grouped in time, a coincidence which was to be repeated time and time again in the future. In this case, as the distinguished historian of Chinese science, Joseph Needham, has shown, some of these experiments had been anticipated in China as long ago as the 2nd century BC in a culture which had and retained a very different attitude from that of the Western world to 'the conquest of nature'.

Guericke showed particularly dramatically in 1654 that the atmosphere was a source of power when he demonstrated a piston and cylinder apparatus. The cylinder was secured in an upright position, its open end uppermost and a rod from the piston protruding. A strong rope attached to the piston rod was passed over a pulley above, and twenty

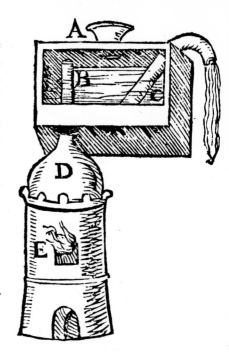

A piece of laboratory apparatus designed by Giambattista della Porta of Naples in 1606 demonstrates the expansive power of steam. A fire heats two ounces of water in a small retort, generating steam which passes into the top of the closed receiver above; the steam then forces water to flow out through the tube on the right.

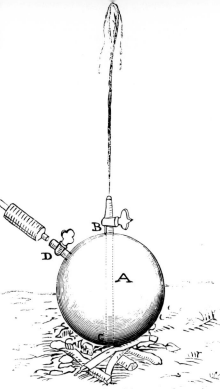

Another demonstration of the power of steam is provided by this sphere which Salomon de Caus, a French landscape gardener, described at about the same time. The sphere is partly filled with water and put on a fire. Some of the water is converted into steam, which forces the remainder up a pipe and out through a nozzle at the top.

men were asked to pull the piston up the cylinder and hold it against the partial vacuum they thus created. Guericke then connected to the base of the cylinder a copper sphere, from which he had pumped out all the air with an air pump of his own invention. As soon as the cocks on the connecting pipe were opened, the residual air in the cylinder rushed into the evacuated sphere with such power that the piston was driven down and the twenty men overcome. By the same method Guericke also succeeded in raising a weight of more than a ton.

From these experiments, well known to the British man of science, Robert Boyle, who carried them further, it was clear that a powerful engine, independent of human, animal, water or wind power, could be built if a vacuum could be obtained under the piston in some other way than by an air pump. In a search for alternatives during the 1670s the Dutch astronomer Christiaan Huygens tried using explosions of gunpowder to expel air from a cylinder, and a French engineer, Jean de Hautefeuille, proposed at about the same time to use gunpowder to power pumps at the Palace of Versailles. But Huygens' idea of using explosions not only for 'violent action' but 'to serve all purposes . . . where man or animal power was needed' – the basis of the internal combustion engine – had a very long way still to go, and in the meantime steam had provided an answer.

Huygens' assistant, a key figure in the story of steam, was Denis Papin, a young Frenchman from Blois, trained as a doctor but just starting out on an energetic career in scientific experiment. Papin was a Huguenot, and in 1675, prompted by the growing menace of religious persecution in France, he moved to London, where he became a Fellow of the Royal Society and read more than a hundred papers to his colleagues. It was in London, too, that Papin gave demonstrations of a double-acting air pump (developed in Boyle's laboratory where he worked) and of 'a new Digester and Engine for softening Bones' – identifiable in retrospect as the world's first steam pressure-cooker.

Papin lived later in Venice, Marburg and Cassel, producing many ingenious devices that worked by vacuum. Following in Huygens' footsteps, he continued to experiment with the gunpowder engine. There was a puzzle here too, since he always found that 'after the flame of the gunpowder' was extinguished 'about a fifth part of the air' remained. Papin turned for guidance both to men of power and to philosophers like Gottfried Leibnitz. In his search for an agent that would leave no residue of air, he chose eventually – and it was a decisive step – to employ nothing but water and steam.

In an interesting letter of the early 1690s, Papin explained clearly just what he had in mind. 'As water has the property that it is converted by fire into steam . . . and can then be easily condensed again by cold, I thought it should not be too difficult to build engines in which, by means of moderate heat and the use of only a little water, that complete vacuum could be produced which had been sought in vain by the use of gunpowder.'

As an experiment, Papin took a small brass cylinder two and a half inches in diameter, fitted it with a piston, and boiled a little water in the bottom. The steam raised the piston to the top of the cylinder, where it was held by a catch. Papin removed the fire and allowed the steam to condense, creating a partial vacuum. When he released the catch, the piston went down with sufficient force to raise by a cord and pulley a weight of sixty pounds.

Papin considered that such a steam engine could be used both for raising water and for other economic and social purposes. He argued, moreover, that it would have distinct advantages over water power. To those – and they were the majority – who suggested that 'the power of rivers' would be more effective (as in the great waterworks at Marly in France, where many waterwheels provided an impressive pumping display), Papin replied: 'There are many cases where rivers are completely lacking or are so distant that the maintenance of a machine would cost almost as much as if water was raised by human power.'

What prevented immediate progress was the practical difficulty of building steam engines on a large scale – in particular, of boring a

In a dramatic experiment conducted by Otto von Guericke, mayor of Magdeburg, in 1654, twenty men pull up a piston in an open-topped cylinder (separately illustrated, with the piston removed). Von Guericke applies a sphere from which the air has been pumped to a cock at the base of the cylinder. When he opens the communication, air beneath the piston rushes into the sphere, and the pressure of the atmosphere forces the piston down, overpowering the twenty men in an instant.

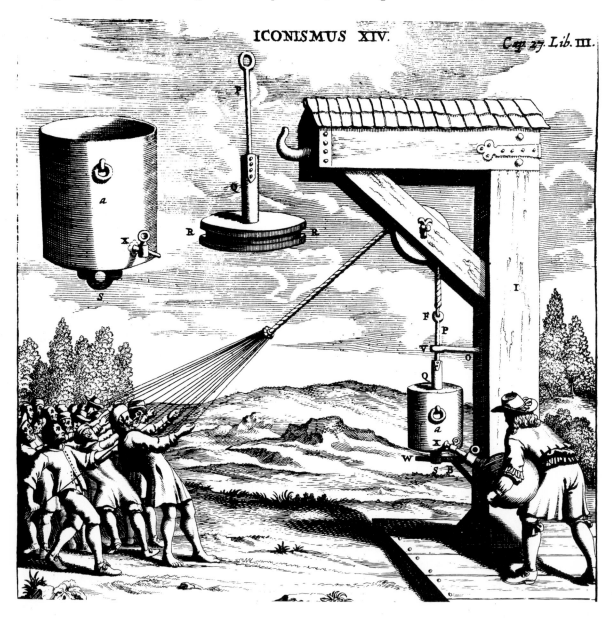

The French scientist Denis Papin, portrayed with an engraving of one of his model engines in 1689, was the first man to operate a piston in a cylinder by means of steam. He allowed steam to cool and form a vacuum under the piston, so enabling atmospheric pressure to drive it down.

large cylinder accurately so that the piston would be a tight fit. In consequence, Papin never progressed beyond the setting up of his laboratory apparatus. Just when he seemed to have reached the brink of practical success, he turned aside to other types of experiment. He found it difficult also to raise financial support, and was given no help by the Royal Society which was then passing through an unenterprising phase. Much that later engineers accomplished was anticipated by Papin, but his promising career ended in frustration and poverty, and he died no one quite knows when or where.

Papin was not the only man of science in the 17th century who was thinking along these lines. In England, the challenge of underground mining was prompting many experiments and proposals for water-raising machines. As early as 1568, Burchard Kranich, a German engineer, had tried devices in Cornwall to 'drain mines and to convey water from any place whatsoever from high to low', and in 1631, when the need was even greater, a Scotsman, David Ramsey, was granted a patent for machines whose purposes speak for themselves:

'To Raise Water from Lowe Pitts by Fire.

To make any sort of Mills to goe on standing waters
by Continual Moc'on without the Help of Windes,
Waite or Horse.

To make Boates, Shippes and Barges
to goe against strong Winde and Tyde.'

Ramsey gave no details of how such machines were to work, and there is no evidence that any were built. But the Marquis of Worcester, a rather self-important inventor, took up Ramsey's themes and built some machines for which he made sweeping claims.

In 1663, the Marquis, whose name figures in all histories of the steam engine, secured monopoly rights for ninety-nine years over a 'water-commanding Engine by him invented', and he, rather than Papin, was regarded as the originator of the steam engine by both Henry Beighton, who drew the earliest known illustration of a steam engine, and by James Watt, who in 1792 described the early history of the steam engine very simply indeed, as follows: 'The Steam Engine, invented by the Marquis of Worcester, was brought into use, in an imperfect state, by Captain Savery. Mr Newcomen added to it a Piston, moving in a Cylinder, and working a detached Pump; and thereby greatly changed it's [sic] Mode of Action as well as reduced it's Consumption of Fuel.' This was a very English pedigree, supported by the Reverend Dionysius Lardner, who proudly described the steam engine as 'the exclusive offspring of British genius fostered and supported by British capital'.

As in most pedigrees, it is the early links which are most debatable. The Marquis of Worcester's flair for self-advertisement was great, perhaps greater than his practical achievements. Yet in a pamphlet published in 1663 with the imposing title *A Century of the Names and Scantlings of the Marquis of Worcester's Inventions*, in which his Invention No. 68 was introduced – an apparatus 'to drive up water by fire' – he claimed to have seen water run from the machine 'like a constant Fountaine-stream forty foot high'. 'One Vessel of water rarefied by fire,' he wrote, 'driveth up forty of cold water.' Judging from this description, which is obscurely worded (perhaps intentionally), such a machine could have been a large version of della Porta's experimental water pump of 1606, except that in place of a single sealed container it had two receivers, apparently made from the barrel of a cannon, which could be filled and fired alternately, so providing a continuous operation. The Marquis maintained that one man could turn the two cocks and at the same time 'abundantly perform' the task of tending the fire.

Another Worcester 'invention', No. 98, was described no less obscurely than No. 68 as 'an engine so contrived, that working the *Primum mobile* forward or backward, upward or downward, yet the pretended operation continueth'. But the 'force of fire' is not mentioned, and although many people have suggested that this was a steam engine, the belief appears to be ill-founded.

Claims to have been the originator of the steam engine have also been made for the lesser known Sir Samuel Morland, 'Master of Mechanicks' to Charles II, who was granted a patent by the King for an invention 'for raising any quantity of water to any height by the help of fire alone'. The King saw several trials of Morland's machine and declared himself 'fully satisfied that [the invention] was altogether new and may be of great use for the clering of all sortes of mines, and also applicable to divers kinds of manufactures within our dominions'. Unfortunately the design of this engine, too, is only conjectural, although if it had cylinders and pistons, worked by steam under pressure from a separate boiler, as some believe, it would have been the first engine of this kind in the world. Morland, like Papin, ended his days in poverty, after failing to provide efficient waterworks for Louis XIV's imposing new palace at Versailles, and his achievement in the history of steam power remains shrouded in mystery.

Whatever precisely were the advances of engineers in the reign of Charles II, one thing at least was clear: the future would be radically different from the past. As Joseph Glanvill, a Fellow of the Royal Society and Chaplain-in-Ordinary for the King, prophesied: 'I doubt not but posterity will find many things, that are now but Rumours, verified into practical Realities.' Glanvill made this prophecy about the coming of the technological age almost a century before the word technology was invented.

The first man to produce and sell a workable steam apparatus for raising water figured prominently in Watt's simple account of the history. Thomas Savery, who died in 1715, was the most prolific inventor of his time, hymned in 1754 by a poet who wrote verses in the most enthusiastic terms:

> 'SAGACIOUS Savery! Taught by thee
> Discordant elements agree,
> Fire, water, air, heat, cold unite,
> And listed in one service fight,
> Pure streams to thirsty cities send,
> Or deepest mines from floods defend.
> Man's richest gift thy work will shine;
> Rome's aqueducts were poor to thine!'

Horse-power and steam-power work side by side at a pithead painted in about 1820. An uncovered Newcomen engine with a spherical-topped boiler winds coal from two shafts, while another pumps from within a tall engine house (left).

'Coming from the Mill', painted by L. S. Lowry in 1930, captures the second phase of the steam age: the use of steam engines, starting in the 1780s, to drive factory machinery. Lowry, who lived all his life in South-East Lancashire, was one of the first painters to make industrial life the inspiration for his art.

Under the canopy of the Crystal Palace, visitors to the Great Exhibition of 1851 in Hyde Park, London, examine exhibits in the Moving Machinery section.

Among the many steam-powered machines on show is Sir Joseph Whitworth's Patent Self-Acting Duplex Railway Wheel Turning Lathe (right foreground).

Steam escapes from numerous flanges and joints of an old horizontal tandem compound engine, a 19th-century refinement of Newcomen's original idea. This engine was installed by Hargreaves & Co. of Bolton in 1883 to drive machinery in a sanitary pipe works in Yorkshire, and was still working more than ninety years later.

To describe Savery's achievement is easier than to outline his elusive but striking career. He had a Devonshire background, and may have been a merchant or a military engineer. In later life he was generally known as Captain Savery, though no one knows for certain why – some have said he was a sea captain, others that he acquired the title in the tin- and copper-mining county of Cornwall, where engineers were often referred to as 'Captain' – and he held Court appointments from 1705, the year when he became a Fellow of the Royal Society, to 1713, when he seems to have been in financial difficulties.

Savery took out many patents, the most interesting of which was Patent 356 in 1698 for 'a new Invention for Raiseing of Water and occasioning Motion to all sorts of Mill Work by the Impellent Force of Fire, which will be of great Use and Advantage for Drayning Mines, Serveing Towns with Water, and for the Working of all Sort of Mills where they have not the Benefitt of Water nor Constant Windes'. The patent was granted for fourteen years. A year later it was extended for a further twenty-one years by Act of Parliament on the grounds that Savery deserved 'a better encouragement by having a longer term of years allowed him for the sole use and benefit of the said invention'.

This remarkable late 17th-century patent sets the scene for the next century, when steam power was to be widely brought into use for the first time, and when the whole question of patents was to assume major importance. The great German poet Goethe, who was keenly interested in the development of science, noted how the English patent law, following the Statute of Monopolies of 1624, encouraged inventors to treat invention as a source of 'real possessions' instead of 'personal fame'; and certainly 'sagacious' Savery conceived of his own patent in this way. The fact that it was couched in general terms was to prove financially advantageous, but later became a source of argument. How far an inventor should be allowed to secure a patent on the basis of a statement of principle and how far he should be required to give detailed specifications was much debated. The prevailing law failed to provide invulnerable protection, with the result that many patent-holders had to bring lawsuits against infringers. On the other hand, critics doubted whether patents of long duration like Savery's were in the public interest since, if unassailed, they tended to stifle development.

We know little about the origins of Savery's interest in steam except gossip and what he wrote himself in 1702 in his book *The Miner's Friend, or an Engine to Raise Water by Fire Described*. This publication appeared in 1702, a year when 'projects' of all kinds were being proposed and advertised for public subscription. Savery was familiar with waterworks – in 1714 he was to become Surveyor of the Waterworks at Hampton Court – and it was in this capacity, he wrote, that he learned of 'rarefaction by fire' and was encouraged 'to invent engines to work by this new force'. A variety of rumours were put about, however, to the effect that he was influenced by Papin, or more maliciously, that his engine was a copy of the Marquis of Worcester's, and that he had bought up all the Marquis's books he could find in Paternoster Row, then the publishing centre of London, and burnt them in order to conceal the origin of his ideas. Another story, as remarkable as that of James Watt's kettle (*see below, page 50*), was that Savery had 'found out the Power of Steam by chance' and to cover it up invented the tale of how he

plunged a wine glass filled with steam into a basin of water and observed
how water was driven up into the glass by the pressure of the air.

Whether or not there was any truth in these rumours, Savery successfully demonstrated a working model of his steam pump to the Royal
Society in 1699. It was backed by a drawing and described fully in the
Philosophical Transactions of the society for that year. The minute of the
meeting ends with the words 'the experiment succeeded according to
expectation, and to their satisfaction'.

This was the first time a steam-power engine had been successfully
demonstrated to a learned society. The engine, which had no moving
parts except for hand-operated steam and water cocks, was basically a
steam pump operating on the principle described by della Porta. Steam
from a boiler forced water out of a closed receiver and up an ascending
delivery pipe or rising main. In this respect it was a steam pressure
apparatus, like the Marquis of Worcester's. However – and this was a
distinct advance on any previous engine – it also used atmospheric
pressure. In order to draw water up from below, steam was admitted
in the first instance to the empty receiver and condensed by a douche
of cold water on the outside. As a vacuum formed, atmospheric pressure
drove water up a suction pipe and into the receiver, ready to be forced
out again and up the delivery pipe when steam was next admitted from
the boiler. Once water had been delivered to a reservoir above, the
cycle began again.

In one of Savery's early engines, installed at Campden House in
Kensington to raise water for domestic purposes, the cycle could be
repeated four times a minute by turning the appropriate cocks, which
were made of brass, as was the pipework. The receiver – made, like the
boiler, of beaten copper – held thirteen gallons of water; the height of
the suction pipe was sixteen feet; and the height of the delivery pipe
was forty-two feet – a lift of fifty-eight feet in all. The useful work done
by the engine was (to use a later unit of measurement) one horse-power.
Later Savery engines had two receivers, one being filled by suction
while the other was discharged by pressure, so that water could be
delivered continuously.

Savery, who was anxious to prove that he was no mere 'projector', went
into production, setting up the world's first workshop for manufacturing
steam engines in Salisbury Court, London, in 1702. 'For draining of
mines and coal-pits,' he had explained in *The Miner's Friend*, 'the use of
the engine will sufficiently recommend itself in raising water so easy
and cheap, and I do not doubt but that in a few years it will be a means
of making our mining trade, which is no small part of the wealth of this
kingdom, double if not treble what it now is.' An advertisement for his
engines stressed that they were less expensive than 'any other force of
Horse or Hands, and less subject to repair'.

At last the steam engine had left the scientist's laboratory. Salisbury
Court was situated just off Fleet Street, already the centre of London's
Press, where editors were saluting the spirit of invention, and Savery
was full of optimism. His fifty-eight foot water lift at Campden House,
where 3,000 gallons were raised in an hour, was exciting much admiration, as was the ingenious fountain, fed by another of his pumps, at
Lord Chandos's magnificent Syon House. Yet utility concerned Savery

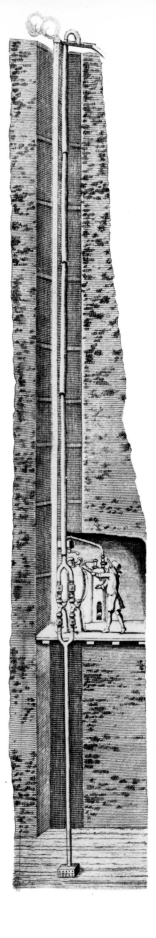

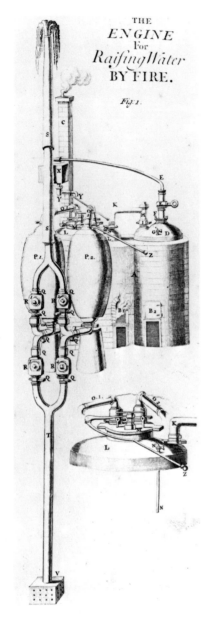

THE
ENGINE
FOR
Raifing Water
BY FIRE.

Fig.1.

The first working steam pump, developed by Thomas Savery and illustrated in his Miner's Friend *of 1702 (above), had two boilers and two receivers, one of which was filled from below by a steam-induced vacuum while the other was emptied to a higher level by steam under pressure, so delivering a continuous flow of water. Savery built several engines to supply large houses, but his scheme to instal them in mine shafts (left) was a failure.*

as much as display. Embarking on his project to drain mines, he delivered some of his engines to mines in Cornwall and Staffordshire, where there was a demand for pumping devices, however experimental. Unfortunately, he never surmounted the technical difficulties inherent in pumping water from depths of 200 or 300 feet, as required in many mines, and after only three years he closed his workshop down and abandoned engine-making on any scale.

Savery's main problem was the same as that which had faced the engineers of Cosimo de Medici. On its suction lift, his pump could not draw water from a depth greater than twenty-five feet; in practice the maximum achieved was even less. Moreover, the low steam pressure obtainable with rudimentary boiler technology and crudely joined pipes restricted the discharge up the rising main to a similar height, especially if a large volume of water was required to be pumped. A total lift of more than fifty feet was rarely achieved. The only way to increase it was to instal engines and boilers at about fifty-foot intervals down mine shafts, an impractical and potentially very dangerous expedient, since Savery boilers were never provided with a safety valve – although Papin had already designed and used such a valve in his 'new Digester'. (The earliest recorded boiler explosion occurred in a Savery engine.) Another shortcoming was the energy loss caused by the fact that during the force lift, steam from the boiler was brought into direct contact with cold water in the receiver.

Savery's most obvious failure was an engine which he installed at York buildings, near the Strand, one of London's chief waterworks, to replace a waterwheel under London Bridge which had previously provided pumping power. This engine gave endless trouble. Its 'strength' was so great, wrote Dr. John Theophilus Desaguliers, another early writer on steam, in his *Experimental Philosophy* in 1744, 'as to blow open several of the Joints of his Machine'. This trouble, caused by using steam at pressures higher than the pipework could stand, was expensive to rectify, and since the engine was never able to pump the quantity of water required, it was taken out of use.

In spite of his failures, Savery remained a pioneer in the eyes of his contemporaries, and he had a considerable band of followers and imitators. One of them, who claimed to be more original than Savery himself, was Dr. Desaguliers. He created variants of the Savery pump in which condensation in the receiver was achieved by an internal injection of water instead of a cold douche on the outside walls. Even Papin, who had made such progress with the cylinder and piston, started to experiment with Savery-type apparatus.

The work continued well into the 18th century. During the 1760s Joshua Wrigley of Manchester was making and selling modified Savery engines, and as late as 1776 – the year of the American Declaration of Independence, and one year after Watt and Boulton secured their twenty-five year patent rights (*see below, page 54*) – William Blakey produced a completely automatic version of the Savery engine. The most successful Savery engines were those which used only the suction part of the cycle – thus avoiding the problems of controlling high-pressure steam. At the end of the century Savery engines were still being used in a flax-spinning mill in Leeds and in Lancashire cotton mills to raise water for waterwheels used to drive the looms, and as late as 1820 a Savery engine was employed to work a waterwheel turning machinery in an engineering works in Kentish Town in London.

The other chief figure in the story of the steam engine before Watt, as Watt himself freely acknowledged, was Thomas Newcomen. By the end of the first decade of the 18th century Newcomen had already captured the centre of the stage. It is a measure of his importance that Britain's leading society for the study of the history of engineering and technology, the Newcomen Society, was named after him when it was founded in 1921. Newcomen's engine was the first practical cylinder and piston engine and if it was, as the 19th-century writer R. H. Thurston described it, 'a combination of earlier ideas', it was also the fore-runner of all subsequent steam engines – indeed of all cylinder and piston engines. It was also the first steam engine ever to be made and erected outside Britain. An installation at Jemeppe-sur-Meuse, near Liege in Belgium, won this honour for Newcomen in 1720–21.

Unlike Savery, Newcomen had no aristocratic or royal connections. Born at Dartmouth in Devonshire, the county with which Savery was associated, he was a Baptist by religion and an iron master by occupation, with an intimate knowledge of the day-by-day pumping problems in tin mines. His assistant, John Calley, was a plumber and glazier. Newcomen never wrote a book, and only one full document survives in his own handwriting.

Exactly how he acquired the ideas to succeed where others had failed is uncertain. It seems unlikely that the details of Papin's scientific work in the *Philosophical Transactions* of the Royal Society could have filtered down to this iron master in rural Devon, and if Newcomen met Savery, as various witnesses reported, the meeting may well have taken place after he had invented his new engine. Marten Triewald, a young Swedish engineer who met Newcomen and had a working knowledge of his engine – he subsequently erected one himself in Sweden – wrote that 'Thomas Newcomen, without any knowledge whatever of the specula-tions of Captain Savery, had at the same time made up his mind, in conjunction with his assistant, a plumber by the name of Calley, to invent a fire machine for drawing water from the mines'. Another early scientific writer, Stephen Switzer, said he was 'well informed that Mr. Newcomen was as early in his invention as Mr. Savery was in his'.

Whatever the origin of his ideas, Newcomen probably spent at least twelve years experimenting with models before he built a full-scale steam engine. The first such engine may have been installed at the Wheal Vor ('Great Work') tin mine in Breage, in Cornwall in 1710, though there is a vague tradition of an even earlier engine being erected at Wendron not far away. But it is unlikely that these machines were more than partially successful.

It is only in 1712 that we turn from conjecture to fact, although even this body of fact has had to be accumulated laboriously by Newcomen enthusiasts. The first recorded Newcomen engine was built in that year – to drain a colliery at Tipton, close to the prominent landmark of Dudley Castle near Wolverhampton in Staffordshire. Although there have been arguments as to where exactly the engine was sited, an engraving of it published in Birmingham by Thomas Barney, a file manufacturer, in 1719, provided a detailed picture of its working parts.

Following a study of the engraving, the historian L. T. C. Rolt emphasized in 1963 what a 'stupendous advance' this engine represented on anything earlier. 'In its purposeful and ordered complexity it makes Savery's crude pump or the toys of Papin and his fellows so primitive by comparison that it is scarcely credible that so few years separate them.'

The engine consisted of a brass cylinder and piston, set up vertically like Guericke's with the top of the cylinder open to the air. Steam was fed in under the piston from a separate boiler and was condensed by a jet of cold water which played inside the cylinder. A vacuum formed beneath the piston, allowing it to be driven down by the pressure of the atmosphere acting on the top, so producing the power stroke. The piston rod was connected by a chain to one end of a wooden beam, pivoted like a see-saw; the other end of the beam operated the piston of

The earliest known illustration of a Newcomen pumping engine, rediscovered in Worcester College, Oxford, in 1925, was drawn in 1717 by the engineer and scientist Henry Beighton. The engine is probably one erected at Oxclose Colliery in County Durham by Beighton himself, who has drawn and labelled every feature, including the spherical-topped boiler, the cylinder, valve gear and wooden beam, and also a man admiring the pump rods working in the shaft.

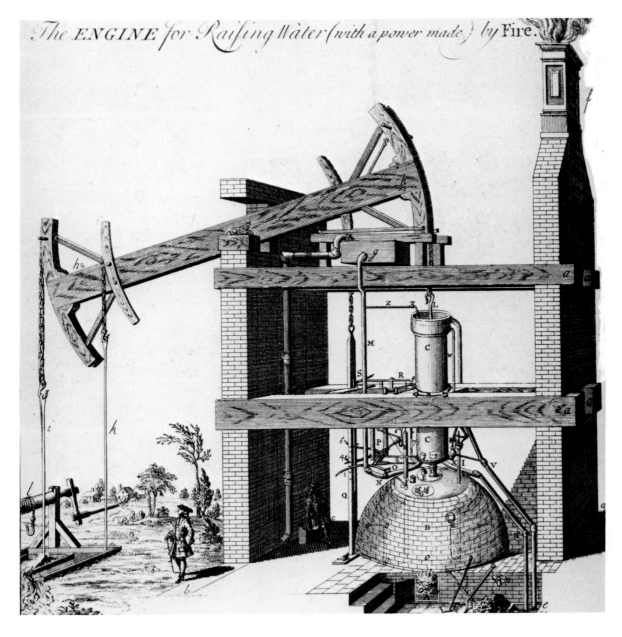

The ENGINE for Raising Water (with a power made) by Fire.

the water pump in the mine shaft. The engine was capable of making about eight or ten strokes a minute and was self-acting – that is, controlled automatically by its own valve gear.

Unlike Savery, Newcomen did not make use of the expansion of steam in his engine. Steam was employed merely to create the vacuum in the cylinder, leaving the work to be done by atmospheric pressure. For this reason engineers often call Newcomen's apparatus an atmospheric rather than a steam engine. Yet Newcomen had an immediate advantage over Savery in consequence, since he did not require to contain and control high-pressure steam; in fact, he used steam at barely more than atmospheric pressure. 'Newcomen hit on the right solution,' Rolt has argued, 'precisely because he was a practical craftsman and so evolved a machine whose construction was strictly within the competence of the craftsmen of the day.'

There was no monopoly of skill, for many ingenious workmen contributed to the successful manufacture of the engine. There was a monopoly of patent, however, which must have been a great disappointment to Newcomen. Because Newcomen created an engine quite different from Savery's, he should have been able to take out a new patent of his own. But Savery's patent, which was general enough to cover any means of raising water 'by the impellent force of fire', prevented him from doing so, nor could he have erected and operated steam engines without infringing it. This impasse was the basis of an arrangement between the two men, probably reached in 1705 (the year, significantly, when Savery closed down his workshop off Fleet Street).

The details of the arrangement remain uncertain, but soon after Savery died in 1715, the master patent passed into the hands of a 'Committee of Proprietors', businessmen based in London who probably secured more of the returns on Newcomen's invention than Newcomen himself. Newcomen also suffered a certain loss of recognition, since the public either associated his name with that of Savery (Thomas Barney's engraving of 1719, for example, described the Dudley Castle engine as 'invented by Capt. Savery and Mr. Newcomen and erected by ye latter 1712') or, in advertisements placed by the patent proprietors, did not learn of his name at all.

The way in which the steam engine patent was exploited was controversial among clients, who in addition to high installation costs had to pay a royalty of around £150 to the Committee of Proprietors in London in return for a licence to erect an engine. The complaints were often noisy. Yet whatever the rights and wrongs, Savery's master patent continued in force until 1733, eighteen years after Savery's death and four years after Newcomen's.

Newcomen's first steam engine of 1712 worked until at least 1725, when, as a diarist recorded, it was still throwing up '60 hogsheads of water in an hour's time' (its maximum recorded output was 120 hogsheads, equivalent to about $5\frac{1}{2}$ horsepower). By then many other Newcomen engines were at work in England and abroad. The West, Newcomen's own part of the country, was as yet poorly represented: in Devon, where tin mining had slumped disastrously, no engines had been built, and in Cornwall the mine captains were still sceptical of what they considered to be a new-fangled machine. Possibly at first they were

Thomas Savery, born into a well-known Devonshire family in about 1650, was one of the most enterprising inventors of his time, and spent considerable sums of his own money on developing his steam pump. He obtained his valuable master patent of 1698 after demonstrating a model of the pump to King William III at Hampton Court Palace.

A table worked out by Henry Beighton in 1721 lists the diameter of pump barrel and steam engine cylinder needed to pump a given quantity of water from a given depth, assuming a six-foot stroke and a speed of sixteen strokes a minute. Beighton was the first man to introduce scientific precision into engineers' calculations, which were then little better than guesswork.

A *Physico-Mechanical* Calculation of the Power of an *Engine*.

Diamerer of the Pump inches.	Draws at a 6 Foot ſtroke.	at 16 St. in a Min. draws per Hour.	The Depth to be Drawn in Yards. (The Diameter of the Cylinder.)												
Inch	Ale Gall.	Hogſh. Gall.	15	20	25	30	35	40	45	50	60	70	80	90	100
4	3. 20	48. 51			Inches		9	10	11	11½	12	13½	14	15	16
4½	4. 04	60. 60				10	11	11¾	12	13	14	15	16	17	18¼
5	5. 02	66. 61			10	11	11¾	13	13¾	14	15½	16¾	18½	19½	20¼
5½	6. 26	94. 30		10	11	12	13	14	15	15¾	17	19	20	21	22¼
6	7. 22	110. 1	9½	11	12	13	14	15½	16	17	19	20½	22	23	24½
6½	8. 46	128. 54	10	12	13	14	15½	16¼	18	19	20	22	23	24½	26½
7	9. 82	149. 40	10¾	13	14	15½	16¾	18¼	19	20½	22	24	25½	27	28½
7½	11. 32	172. 30	11	13¾	15	16½	18	19	20	21¼	23¼	25	27	28½	30¼
7¾	12. 02	182. 12	12	14	15½	17½	18¾	19¾	21	22	24¼	26	28	29	31½
8	12. 82	195. 22	12½	14½	16½	18½	19	20½	21½	23	25	27	29	30½	32½
8½	14. 52	221. 15	13½	15¼	17¼	19	20¼	21¾	23	24	26¾	28½	31	32½	35¼
9	16. 24	247. 7	14	16¼	18	20	21½	23	24¼	25	28	30½	33	35	36½
10	20. 04	304. 48	15½	18	20	22	23¾	25¼	27	28¾	31¼	33¾	36	38¼	40

deterred by the indifferent performance of Newcomen's earliest experimental engines; later, they would certainly have been put off by the enormous consumption of coal, a commodity which had to be imported into the West Country. Newcomen's early successes came in colliery districts where high coal consumption was not such an obstacle: in the Midlands, where, as Rolt interestingly showed, his Baptist connections aided sales, and in the North Country, where the first engine was installed near Leeds in 1714–15.

There were two early Newcomen engines at Griff, near Nuneaton in Warwickshire – the first of them built in 1714, with features that were much copied in later engines – and an engine erected at Whitehaven in Cumberland in 1715, which was ranked by an interested observer as 'above all others that I have yet heard of for ingenious contrivance'. In 1719 the first engine in Lancashire was installed to drain coal mines near Prescot, and in 1726 a Newcomen engine was supplied to York Buildings in London to replace the unsuccessful Savery engine. The first steam engine in Wales was erected at Hawarden in Flintshire in 1714 by Richard Beach, a member of a coal- and iron-mining family, who became a great advocate of the steam engine and is the first person known to have used cast-iron cylinders and pistons.

The cost of iron was about one third that of brass, the metal previously used. Yet the switch-over, which greatly encouraged the building of engines, would have been impossible had it not been for that key innovation in industrial history, the introduction of iron-smelting with coke, first achieved by Abraham Darby at Coalbrookdale in Shropshire in 1709. By the time the Savery/Newcomen patent expired in 1733, the number of iron cylinders cast by Darby's foundry at Coalbrookdale had risen to twenty-two – the total for iron-cylindered steam engines in existence, since as yet no other foundry was undertaking such work.

Authority to develop the Newcomen engine in Scotland was granted by the Committee of Proprietors in London to various engine builders from 1719 onwards. One engine at a colliery at Elphinstone in East Lothian, built in 1720, had a cylinder with a twenty-eight-inch bore and replaced no less than fifty horses which had previously been employed to power the drainage pumps. By 1800 more than twenty-five Newcomen engines were said to be at work in the Ayrshire coalfield alone.

Abroad the Newcomen engine, extolled by a German observer as

'the beautifullest and most perfect engine that any Age or Country ever yet produced', was introduced in 1722 into Hungary (where an installation was begun at Königsberg in the mining district of Upper Hungary, now part of Czechoslovakia), into Germany (at Kassel), and in 1724 into Spain (at Toledo). The Königsberg engine was in full operation by 1724, and there is a 1753 drawing of it in the Deutsches Museum in Munich. This project had been instigated by Emanuel Fischer von Erlach, son of the baroque architect Johann Fischer von Erlach, who worked at the Court of Vienna. The younger von Erlach had seen the new power in action while on a visit to England, and introduced a number of improvements, some of which were incorporated in 1722–23 in an engine with a seventy-five-foot lift to drive the fountains in the gardens of Prince von Schwarzenberg in Vienna. Frederick the Great of Prussia was interested in such developments and had drawings of the new engines made for his library.

In 1726, with considerable publicity, an engine was installed at Passy near Paris to raise water from the River Seine for the growing French capital, and a year later an engine built by Marten Triewald was installed at the Dannemora mines in Sweden; its thirty-six-inch brass cylinder, cast at the Imperial Gun Foundry in Vienna, was then the largest in existence. Such foreign ventures publicized English 'invention' and earned mentions in publications such as M. Belidor's *Architecture Hydraulique* of 1724, which stated that the manufacture of 'fire engines' was exclusively confined to England.

For all these successes, when Newcomen died in 1729 he was still virtually unknown outside a small circle of engineers. Indeed, although a brief obituary in the *Monthly Chronicle* honoured him as 'the sole inventor of that surprising machine for raising water by fire', his death

All kinds of fanciful images were littered through the pages of 19th-century steam engine histories. Among terse engineering prose and mechanical diagrams, cherubs flew to the moon on chariots of fire, babies rocked in patent steam cradles, steam cannon were fired and bottoms were burnt by steam from boiling kettles. In this way historians such as Robert Stuart and R. W. Thurston revealed the lively influence still exerted by the steam toys of the classical mechanicians as well as the wilder reaches of the contemporary imagination.

received little public attention. It was not until 1921 that a memorial to him was built at Dartmouth, his birthplace. More than forty years later still, in 1963, a memorial engine was erected to celebrate the three-hundredth anniversary of his birth.

The steam engine itself, however, was before he died already receiving ample praise in the somewhat extravagant language of early 18th-century writers. Many poems were to be written about the steam engine, but none were so remarkable as the *Aenigma*, published in 1721, when the characteristics of engines were still ranked among the more arcane mysteries of natural philosophy. As its title indicates, it was in the form of a riddle for leisured readers to solve, and it appeared in a most unlikely journal, the *Ladies Diary*, whose editor was the energetic Henry Beighton. (It was he who was responsible for a table of data, published in the same journal, which set out the engine specifications required to pump a particular quantity of water from a particular depth.) Beighton was probably also the author of the poem, in which classical allusions and a knowledge of anatomy and physiology were extensively displayed:

'I sprung, like Pallas, from a fruitful Brain,
About the Time of Charles the Second's Reign,
My Father [the Marquis of Worcester] had a num'rous Progeny
And therefore took but little Care of me . . .
Yet sending Pictures of my Face to draw –
And many of his other Children too . . .
He then to me strange Education gave,
Scorch'd me with Heat, and cool'd me with a Wave:
More Work expected from my single Force,
Than ever was performed by *Man* or *Horse* . . .
To mend my Shape, he oft deform'd it more
Which sometimes made me Burst, and Fret, and Roar . . .
My heart has *Ventricles*, and twice three Valves;
Tho' but one *Ventricle*, when made by Halves.
My *Vena Cava*, from my farther Ends
Sucks in, what upward my great Artery fends,
The Ventricles receive my pallid Blood,
Alternate – and alternate yield the Flood;
By VULCAN's Art my ample Belly's made –
My Belly gives the Chyle [the steam] with which I'm fed –
From NEPTUNE brought, prepar'd by VULCAN's Aid . . .
On mighty Arms, alternately I bear
Prodigious weights of Water and of Air;
And yet you'll stop my Motion with a Hair.'

Anatomy and physiology, themselves still young subjects, provided a fertile source of ideas, not only in the *Aenigma* but in the prose writing of the time and in captions to engravings. For example, in a list of engine parts referred to as 'the several Members' in Thomas Barney's 1719 engraving, one finds items such as 'The Fire mouth under the Boyler with a Lid or Door' or 'The Neck or Throat betwixt the Boyler and the Great Cylinder'. In this kind of vocabulary the novelty of the steam engine, so suddenly invading the early 18th-century world, is eloquently captured. Few people then can have imagined how common-place 'the Fires and Boylers' of steam engines were soon to become.

At a South Wales colliery steam billows from
two engine houses, each driving the winding
gear for a separate shaft. Around 1900, when
this photograph was taken, a third shaft with a
high bricked top (right) provided updraught
ventilation by means of a furnace at the bottom
of the pit and a steam-driven fan.

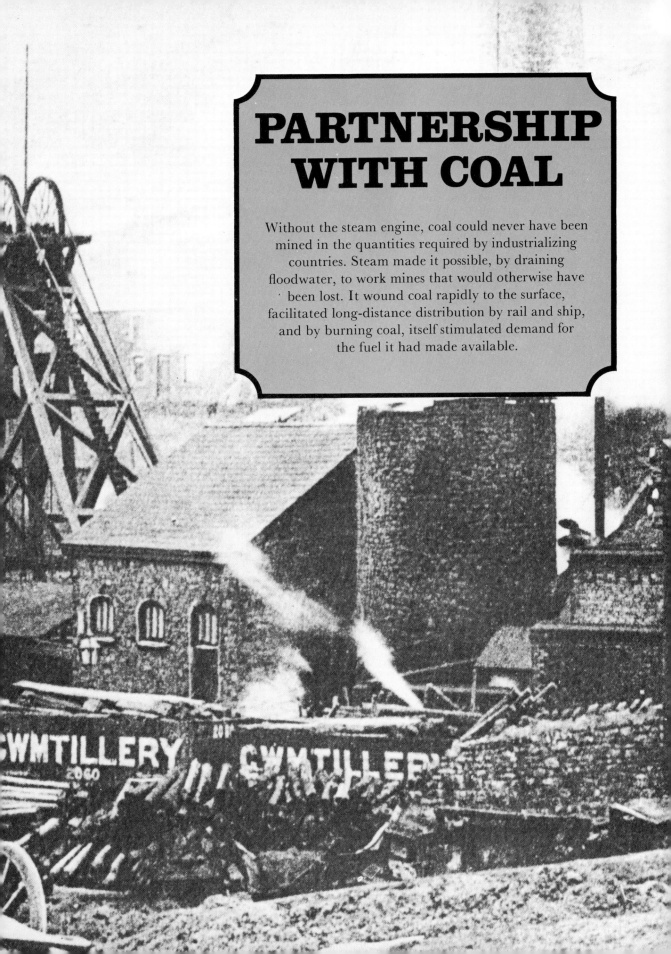

PARTNERSHIP WITH COAL

Without the steam engine, coal could never have been mined in the quantities required by industrializing countries. Steam made it possible, by draining floodwater, to work mines that would otherwise have been lost. It wound coal rapidly to the surface, facilitated long-distance distribution by rail and ship, and by burning coal, itself stimulated demand for the fuel it had made available.

Stokers fuel a battery of egg-ended boilers at a Lancashire colliery. The coal is carried into the fires mechanically on iron conveyor belts.

Pithead Slums

Colliery boiler houses were the slums of the boiler-house world – filthy with coal dust and usually open to the weather. Big pits had up to twenty boilers, often fitted with mechanical stokers to reduce the effort of fuelling a furnace fifteen feet long.

Brick arches cover the 70-year-old boilers, still in use in 1973, at Morlais Colliery, Dyfed.

In the Morlais firing aisle, gauges record the steam pressure and water level of a Lancashire boiler.

A driver of 1890 brakes the winding engine at Rose Bridge Colliery, Wigan. In front of him are dials similar to those below, on which moving hands record instructions from the 'onsetter' in the pit and indicate the positions of the cages as they are raised or lowered in the shaft.

A twin-cylinder horizontal winder of 1893, still running in the 1960s, pauses between winds. The second of the two cylinders is by the driver (far left).

Ropes and chains festoon the pumping gear at a Newcastle pit top, where a 'sinker' in cape and rimmed hat has been at work on a new seam, opened in 1892.

Underground Traffic

By the late-19th century many pits had boilers and engines underground as well as winding gear powered from the surface. One pit in the north of England even had boilers in a seam, separated from the coal only by thin brick vaults. The exhaust gases fed the draught in the upcast shafts, helping to create winds of up to 40 m.p.h. in the downcast shafts.

Steam gradually superseded pit ponies for long-distance haulage below ground. An engine at a pit near Bristol could haul 40 tubs each holding 9 cwt. of coal over an undulating plane 2,000 yards long.

An onsetter signals to the engine driver at the pithead to wind the loaded coal tubs up the shaft.

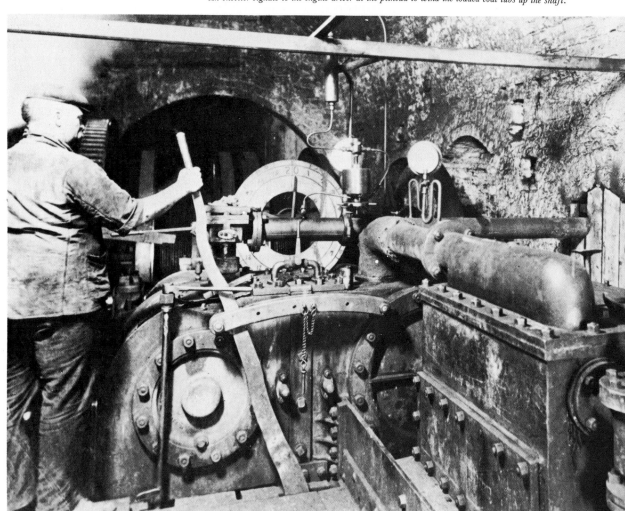

At Frog Lane Colliery near Bristol, a miner stops the underground compound haulage engine. Installed around 1890, it hauled 250 tons of coal a day.

Miners in cages wait for a steam-powered ride.

One leg in and one leg out, North Country shaft sinkers travel four to a bin in a makeshift lift.

A foreman directs surface workers, some no more than boys, others elderly or injured, as they grade coal on the sorting belts of a North Country pit.

Miners at a Welsh colliery stand by as a Peckett saddle-tank engine prepares to haul away a train of wagons that have been filled at the screens.

Surface Handling

When coal had been mined and wound to the surface, banksmen unloaded it on to conveyor belts running to the jigging screens. There it was sorted and graded by men generally too old and infirm or else too young to be at the coal face – a form of local welfare before the age of the welfare state.

The sorting belts were driven by small horizontal engines, the 19th-century maids-of-all-work, or sometimes by old haulage or capstan engines adapted for the purpose. After about 1900, electric motors began to be used, running off current produced by the collieries' own steam-powered generators.

From the screens house coal and steam coal was separately tipped into waiting trucks and shunted by colliery locomotives to the main lines for delivery by rail to the towns and cities.

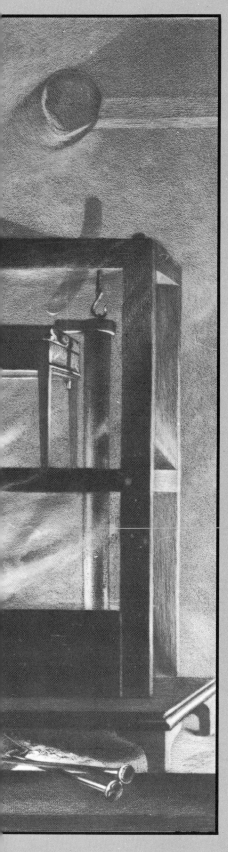

CHAPTER TWO

JAMES WATT AND THE INDUSTRIAL REVOLUTION

Before the era marked by the discoveries of
JAMES WATT, the steam engine, which has since
become an object of such universal interest, was a
machine of extremely limited power, greatly inferior
in importance to most other mechanical contrivances
used as prime movers. But from that time it became
a subject not of British interest only, but one with
which the progress of the human race became
intimately mixed up.

REV. DIONYSIUS LARDNER,
*The Steam Engine, Familiarly Explained and
Illustrated* (1827)

*James Watt studies the model Newcomen
engine whose deficiencies led him in the winter of
1763–64 to invent his separate condenser. By
the time this mezzotint appeared in 1869,
Watt's success had already made him a
legendary figure.*

The name of the Scots engineer James Watt was and still is to many people virtually synonymous with the steam engine, and the mistaken belief that he invented the steam engine long enjoyed popular currency. Watt's reputation rose so high among the Victorians that when the Lord Provost of Glasgow, Sir James Bell, wrote a history of his great city in 1896, he found it a matter of pride to begin with the achievement of Watt.

'According to tradition,' Bell's first paragraph starts, 'it was while taking his accustomed walk on Glasgow Green on a pleasant evening in the spring of 1765 that the idea of the separate condenser to the steam engine flashed across the mind of James Watt. Every circumstance connected with that conception was momentous and memorable. No innovation, in any time or country, was fraught with such far-reaching and revolutionary consequences in the economic and social relations of the human race.'

The separate condenser to which Bell referred was the biggest single improvement ever made to the steam engine. By condensing the steam in a separate chamber rather than in the cylinder itself, Watt was able to create a far more efficient steam engine than Newcomen's. To exploit his invention commercially, he joined the leading Birmingham manufacturer Matthew Boulton in 1774, forming one of the most celebrated partnerships in world economic history. By the year 1800, when patent protection ran out, Boulton and Watt had sold over 470 engines, nearly two thirds of them of a double-acting rotative type. This innovation, the second of Watt's major contributions, enabled steam not only to pump water, as hitherto, but efficiently to drive machines of all kinds, so promoting the revolutionary social and economic consequences that Sir James Bell had in mind when he began his book.

Myths soon grew around Watt. Generations of schoolchildren were regaled with the classroom story that he had discovered the power of steam while watching the lid of a boiling tea kettle as it rattled up and down. In another version of the story, an aunt reproached him for wasting his time by repeatedly taking off and putting on the lid of the tea kettle, holding saucers and spoons over the steam, and trying to catch the drops of water formed on them by vapour. The story was supposed to point a moral about close observation, and it is at least in part a myth justified. Watt may or may not have learned anything from observing a tea kettle boiling as a boy, but he did carry out experiments in Glasgow using a kettle as a small boiler.

It is interesting to note that the year 1765, when Watt took his solitary walk on Glasgow Green, marked the beginning of a run of bad harvests and, as in Savery's day, 'sanguine schemes for reclaiming land, building roads and bridges, and opening manufactories of all sorts' were prevalent; and also that many people beside Watt were working on new kinds of steam engines. Boulton compared them colourfully with 'Tubal-Cains' or 'Dr. Fausts . . . arising with serpents like Moses'. All started from the same point: the engine of Thomas Newcomen. Watt was no exception: he did not begin *de novo*. Like the others he was an improver, but one of exceptional brilliance.

Pumping engines were already in wide use by the 1760s: Dr. Kanefsky and Dr. Robey have recently estimated that 200 to 250 were in service out of a total of 320 to 350 which had been built. After the expiry of the

An aunt looks on with uncomprehending disdain as the boy James Watt, scientist in the making, plays with the boiling tea-kettle. This supposed incident was elevated in Victorian England to the status of a parable.

Savery/Newcomen master patent in 1733, mine owners who had baulked at paying the premium demanded by the patent holders in London more readily abandoned their horse- or water-powered pumps. Engine building was thrown open to any iron masters who cared to offer their products. J. R. Harris, one of the first historians to attempt a careful count of Newcomen engines, estimated in 1967 that whereas a mere sixty engines were constructed between 1712 and 1733, no fewer than 300 were built between 1734 and 1781. The largest concentrations were in Cornwall and the North-East.

As the number of engines increased, striking developments in design and power took place. How much fact and how much legend there is in attributions of particular improvements in the Newcomen engine to particular people is a matter of interest primarily to historians of technology with an antiquarian as well as an archaeological bent, but certain names deserve to be singled out, among them the Hornblowers who were to present Watt with powerful competition in the West Country. With the settling in Cornwall in 1744 of Jonathan Hornblower, son of the Staffordshire engineer Joseph Hornblower, there was an increase in the use and sophistication of engines in Newcomen's home territory, encouraged by the remission, after a long political struggle, of duty on coals consumed in the drainage of mines. By 1758 at least thirteen engines were running there; by 1769 the number had increased to eighteen, of which eight had cylinders larger than sixty inches in diameter, and by the end of the century to 145. It was a Hornblower (Josiah) who in 1753 accompanied the first steam engine delivered to the United States. (What is said to be the lower half of the cylinder is preserved in the Smithsonian Institution in Washington.) Another Hornblower (Jonathan) patented a compound engine in 1781, but the patent was revoked after Watt protested that it was an infringement of his own patent. (The later development of the compound engine is described in Chapter 5.)

In the search for greater power, engineers before Watt had tended to concentrate on increasing the size of the cylinder, and some achieved a more regular supply of steam by installing two or more boilers for each cylinder. In the North-East of England, where there was cheap coal on the spot, large and powerful engines of a type constructed by William Brown became popular. One of them, built in 1763, had a cylinder 76 inches in diameter and was served by four boilers, one of which was kept spare. Another engine of that date in Bristol had a $74\frac{1}{2}$-inch cylinder weighing $6\frac{1}{2}$ tons, and an engine with a 76-inch cylinder was constructed at Wednesbury in Staffordshire in 1775.

Two great engineers whose names figure in all histories of the steam engine sought improvements in different ways, the first marginally, the second more comprehensively. James Brindley, who erected an engine in Staffordshire in 1756, was one of the first to experiment with new kinds of boilers, and John Smeaton, who lived at Austhorpe near Leeds, introduced many changes, some of which doubled engine performance. Smeaton began in 1765 by making an experimental one-horse-power model; it had an oscillating wheel instead of a beam, the boiler had an internal flue (an idea tried by Brindley), and the cylinder was small in diameter in proportion to the length. Smeaton built a full-size engine with some of these features in London in 1767, but it did not measure

As a page from Watt's notebook shows, the famous kettle was more than just a schoolroom story.

up to his expectations. This disappointment led him to undertake a full-scale survey of engines around the country.

Unlike empirical improvers, who solved problems by trial and error, Smeaton carefully computed mechanical and thermal efficiency and was as much concerned with the waste of steam as with its power. He was able to demonstrate beyond doubt that the engine with the largest cylinder was not necessarily the most efficient or even the most powerful. He uncovered all kinds of deficiencies in existing engines, the worst of them often very crude indeed, criticising weaknesses in component parts – bad casting of cylinders, for example, faulty construction of valve gear and wrong placing of grates – and even commenting on the quality of the water used (hard water produced scaling in the boilers, reduced efficiency and increased the risk of boiler explosions).

James Watt, like Smeaton, was concerned about the inefficiency of existing steam engines, and like Smeaton he began with models. In the winter of 1763–64, then in his late twenties, he was asked to repair a model of a Newcomen engine kept in Glasgow University. Though still an obscure mathematical instrument maker, he had made friends at the university with, among others, Dr. Joseph Black, Professor of Chemistry and explorer of latent heat. Watt found the model engine consumed so much steam that it could not run for more than a few strokes at a time. This problem had already defeated a well-known London instrument maker, Jonathan Sisson, but where Sisson had failed, Watt succeeded. By working on the model he not only grasped the principles of the Newcomen engine, but saw how to improve it.

Realizing that steam was wasted by the continual heating, cooling and re-heating of metal as first steam and then cold water to condense it was introduced into the cylinder, he proceeded from first principles, setting out his reasoning with remarkable clarity: 'In order to make the best use of steam, it was necessary – first, that the cylinder should be maintained always as hot as the steam which entered it; and secondly, that when the steam was condensed the water of which it was composed, and the injection itself should be cooled down to 100 degrees, or lower where it was possible . . . In 1765 it occurred to me that if a communication were opened between a cylinder containing steam and another vessel, which was exhausted of air and other fluids, the steam as an elastic fluid, would immediately rush into the empty vessel, and continue to do so until it had established an equilibrium; and that if the vessel were kept very cool by an injection, or otherwise, more steam would continue to enter until the whole was condensed.'

Watt achieved further efficiency by covering the top of the cylinder and using low-pressure steam rather than atmospheric pressure to drive the piston down when a vacuum formed beneath it. He also fitted the cylinder inside an outer steam case so as to maintain its temperature. To scavenge the condensed water and the injection water from the separate condenser he provided an air pump worked off the engine beam. He secured funds for developing his engine from Dr. John Roebuck, a Birmingham man who directed the great Carron Iron Works in Scotland (the second foundry to produce cast-iron cylinders for steam engines). Watt began with little experience either of 'the construction of large machines' or of business demand for them. But by 1769 he had completed a small experimental engine with a

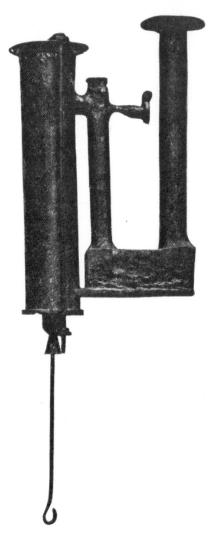

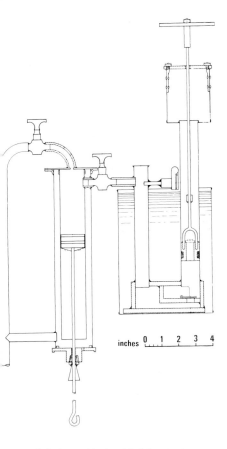

inches 0 1 2 3 4

A tinplate and lead model of the separate condenser engine, the second Watt made (left), is conjecturally drawn in section (above) with missing parts added. From the left, steam passes through the steam cock and enters the inverted cylinder, pushing down the piston. When the cock in the cylinder wall is opened, the steam rushes into the adjoining condenser, which is immersed in a cistern of cold water and scavenged by the air pump on the right. A vacuum rapidly forms in the cylinder, and the piston is drawn up on its power stroke. The cylinder, surrounded by a steam jacket, remains hot all the time – a great advance on the Newcomen engine, whose cylinder was cooled by an injection of cold water, and had to be reheated, at every stroke.

cylinder eighteen inches in diameter and a stroke of five feet, and had taken out his first and most famous patent (No. 913) setting out 'a new method of lessening the consumption of steam and fuel in fire engines'. Writing to a friend, he remarked that he was 'not the same person I was four years ago when I invented the steam engine'.

The translation of Watt's ideas into full-size steam engines was difficult and protracted, and Roebuck, who became financially embarrassed because of the range of his enterprises, could not supply enough money. His bankruptcy in 1773, a year of trade recession, left the way open for Watt's second partnership with Matthew Boulton.

Watt first met Boulton in Birmingham in 1769. Five years later he went to live and work in Birmingham, taking with him his little experimental engine. The history of the steam engine thereafter moved with him from Scotland to the English Midlands. Here Watt was among men who were fascinated by invention. He found congenial company at once in the Birmingham Lunar Society, a group of people who met regularly to discuss most aspects of science and technology; they included businessmen, foremost among them Matthew Boulton, as well as 'philosophers'. Boulton, who had started as a button maker, was an experienced exporter, interested in manufacturing and dealing in 'articles of consequence', whether they were 'things of use' or 'matters of ornament', and his Soho Works, opened just outside the city in 1762, were already perhaps the best known industrial premises in the world – 'Soho! Where Genius and the Arts preside, Europa's Wonder and Britannia's Pride,' wrote a Birmingham poet – and not far away the pioneering ironfounder John Wilkinson introduced a new boring machine without which it would have been impossible to build cylinders of the quality Watt needed.

It is misleading to treat the partnership between Watt and Boulton as one between 'a natural philosopher' on the one hand and an entrepreneur on the other. Watt came to understand business, and Boulton from the outset was no mean philosopher. If in a sense their gifts – or at least their experiences – were complementary, their partnership rested on an interchange of ideas and a harnessing of drive. It began, however, with vision. Roebuck had thought of obtaining a licence for manufacturing Watt engines in three counties in the Midlands. Boulton's horizons were far wider. William Hazlitt was to write later of his 'capacity for affairs – quickness and comprehension united'. In a deservedly famous letter of 1769, written soon after he met Watt, Boulton was both persuasive and frank.

'I was excited by two motivs [*sic*] to offer you my assistance which were love of you and love of a money-getting ingenious project. I presum'd that your Engine would require money, very accurate workmanship, and extensive correspondence, to make it turn out to the best advantage; and that the best means of keeping up the reputation, and doing the invention justice, would be to keep the executive part out of the hands of the multitude of empirical Engineers, who from ignorance, want of experience and want of necessary convenience, would be very liable to produce bad and inaccurate workmanship; all of which deficiencies would affect the reputation of the invention. To remedy which and produce the most profit, my idea was to settle a manufactory near to my own by the side of our Canal, where I would erect all the conveniences necessary for the completion of Engines, and from which Manufactory we would serve all the World with Engines of all sizes. . . . It would not be worth my while to make Engines for three Countys only, but I find it very well worth my while to make for all the World.' This was a clarion call, no less stirring than Boulton's later claim that he sold at Soho what all the world wanted: power.

In 1776, the year of the American Declaration of Independence, Watt produced his first two large engines; one, the so-called 'Parliament Engine' with a fifty-inch cylinder, was put to work at Bloomfield Colliery in Staffordshire, the other with a thirty-eight-inch cylinder was employed at Wilkinson's iron furnace. Both made about fourteen to fifteen strokes a minute, and in Watt's words, were attractive enough to please even captious workmen. Within a few months of their successful installation, brewers were ordering engines too, one 'to raise 15,000 ale gallons 60 feet high'.

Financial success in this field did not come immediately to Boulton and Watt. While Boulton lengthened the order books, Watt continued to toil at research and experiment. The hallmarks of Boulton and Watt's business were superior performance protected by patent and high standards of workmanship – characteristics which required much time, money and effort to sustain.

The first requirement, as Boulton shrewdly recognised, was to extend Watt's patent 'if most profit' was to be secured. In 1775, the year before the two first Watt engines were installed, Boulton with great skill persuaded a Parliamentary Committee to accept a 'Steam Engine Act' which vested in 'James Watt, Engineer, his Executors and Administrators . . . the sole Use and Property of certain Steam Engines' for twenty-five years. Boulton told the Committee that these 'certain Steam Engines' would save '3 or 4 Times the Quantity of Coals' used in 'the Common Fire Engine' and that they would do '4 Times the Work'. As a precedent for the Committee to consider, he produced a copy of the clause from the earlier Act of Parliament granting patent protection to Thomas Savery.

Watt's private comments reveal just how much importance the two men placed on getting the patent extended and how much effort they put into their lobbying. In a letter to his father in 1776 he wrote with relief: 'After a series of various and violent Oppositions I have at last got an Act of Parliament. . . . This affair has been attended with great Expence and Anxiety, and without many friends of great interest I should never have been able to carry it through, as many of the most powerfull people in the House of Commons opposed it.'

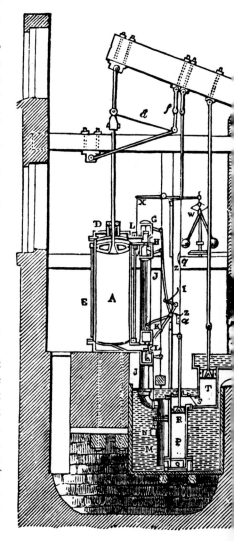

The Watt engines were only partially built at Soho. They were not standardized and depended on many components made locally where the engines were needed. Until 1795, however, Watt insisted that almost all cylinders were bored by Wilkinson, thus ensuring that in the most vital part of the engine high standards of workmanship were maintained. Such standards were high, however, only by contemporary yardsticks. Smeaton doubted whether cylinders could be made with adequate precision for Watt's purpose, and Watt once congratulated himself that one of the Wilkinson cylinders lacked only three-eighths of an inch of being truly cylindrical. The age of precision machine-tools and interchangeable parts was still to come, and the making of steam

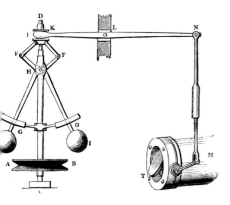

Watt's double-acting rotative engine, illustrated (left) driving a saw mill, was introduced in 1784. It converted the reciprocating motion of the beam to rotative motion by means of sun and planet wheels (marked j and i in the drawing below of the 1788 Albion Flour Mills engine in London). The piston was attached to the other end of the beam by parallel motion gear, and the engine speed was regulated by a governor (above). As the engine picked up, the twin balls moved outwards, pulling down a sliding collar and closing a butterfly throttle in the steam inlet pipe.

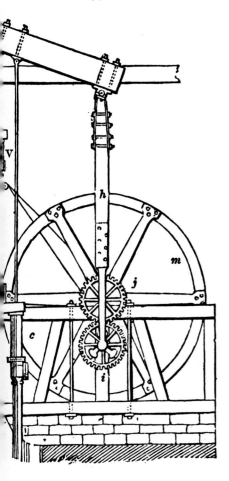

engines was to speed it on. Watt was always properly critical of poor workmanship. 'If possible,' he once wrote to Boulton, 'have the whole brood of these enginemen displaced, if any others can be procured; for nothing but slovenliness, if not malice, is to be expected of them.'

Boulton and Watt's profits came not from the making of the engines but from the royalty or 'premium', as they called it, which they charged on all engines in use. In this they were following the Savery-Newcomen precedent. The amount of the premium was related to the saving in fuel on a Watt engine as compared with a Newcomen engine doing the same work, with Boulton and Watt collecting one third of the value of the saving. Watt engines sold on this basis soon became more widely known than any engines produced by a previous inventor. By 1800 there were said to be fifty-five working in Cornwall alone, although much to the annoyance of the Soho partners this figure represented barely half of all engines in use in the county.

It was a regular source of irritation and of considerable financial loss to Boulton and Watt that in Cornwall they found it difficult to prevent infringement of their patent rights and to collect premium dues. In 1799, therefore, the two partners drew up a revealing financial summary which set out all the key figures in a long-running dispute as they themselves estimated them. It suggested that they had been obliged 'to spend at Law and otherwise upwards of £10,000' to challenge 'leading Adventurers', and that instead of receiving upwards of £300,000 in premiums, as they would have done if the original agreement had been 'strictly enforced and fairly and fully acted upon', they had collected only about £100,000. After making 'Voluntary Concessions' to miners in distress, they claimed arrears of £42,000. Only with great difficulty and several more years' delay were they able to recover most of this sum. Meanwhile, they emphasised, Cornishmen had saved in fuel bills alone from £800,000 to £1 million by using their engines 'in comparison of what would have been required by Common Engines to do the same work during twenty years from 1778 to 1798'.

Economic demands were changing during the last decade of the 18th century, largely as a result of the introduction of new machinery into the textiles industry, and these demands were to produce a corresponding change in the type of engines made by Boulton and Watt, as well as fresh challenges from competitors. Even before the application of steam engines to the driving of factory machinery, in which Watt scored new technical successes, the textiles industry was being revolutionized on the basis of traditional sources of power. Richard Arkwright used horses to power his first cotton mill at Nottingham in 1769, and Edmund Cartwright, inventor of the power loom, employed a bull in a pilot project at Doncaster. Animal power continued to be used even in the early 19th century, surviving longest in the carding branch of the cotton industry, but a more widely used and satisfactory form of power continued to be provided by water.

Arkwright's water frame of 1769 was not quite, as he claimed, the 'new Piece of Machinery never before found out, practised or used', but without Arkwright's enterprise the water-powered revolution in the textiles industry would have been less dramatic. Water power began to be employed more generally during the 1770s, offering great advantages to factories near streams with a reasonable fall and a constant flow. Manufacturers who had easy access to water were not always anxious to change to steam power.

Steam could not be applied directly to cotton spinning and weaving machines until some effective way was found of converting the oscillating motion of the steam engine beam to rotative action, the demand for which attracted many inventors besides Watt. Progress was at first impeded by the belief that the variable stroke of the piston in the cylinder could not be harnessed to a steadily turning wheel. Early engineers, not foreseeing that the wheel itself would impose a regular length of stroke on the piston, introduced what Rolt called 'complex and trouble-fraught combinations of toothed racks, ratchet wheels and gear wheels' in their attempts to make the Newcomen engine rotative.

One of the most interesting pioneers was John Stewart, who wanted to use steam power not in the cotton mills of Lancashire, but in the sugar cane mills of the West Indies. His patent of 1766 bore the simple inscription: 'A Machine which performs its operations by the Power of such Common Fire Engines as are used for raising water out of Mines, which he apprehends will answer all the purposes of Wind, Water and Horse Mills.' A year later in a pamphlet, the only remaining copy of which is in the New York Public Library, Stewart dwelt on the advantages of operating sugar cane mills by steam power. Windmills provided too irregular power, he argued, and at least thirty expensive and stubborn mules were needed in a factory producing two tons of sugar each day. Stewart estimated that the saving to a factory taking up steam power would be £600 a year. One of Stewart's engines was installed in such a factory, but proved a failure.

Until an effective rotative engine was produced, steam was used to enhance water power rather than to supersede it. The limitations of water power were as obvious as those of mules or horses. Since a constant flow was rarely available all the year round, engineers often used Newcomen and Savery engines to pump water back from the tail race to the head race of a waterwheel, so that water could run over the wheel time and time again. One of them was Joseph Oxley, who in 1763 installed two steam engines to assist waterwheels lifting coal from seam to surface at a colliery. Steam engines were also installed to assist waterwheels at Arkwright's mill at Cromford in 1780. Another engineer interested in achieving rotative motion was Smeaton, who preferred 'the smooth torque of the water wheel', as Rolt put it, to the complex systems introduced by men like Stewart.

The breakthrough in converting reciprocating engines to rotative motion by mechanical means came in the 1780s, headed by Watt who was by now at the height of his inventive career. In 1781–82 he made his engine double-acting, as he had long wished to do, by applying steam alternately below and above the piston to produce a power stroke in both directions, so creating an engine more suitable for providing rotative motion. Almost simultaneously, however, a number of other engineers introduced their own rotative engines, notably James Pickard who patented his 'Method of applying steam engines to the turning of Wheels' in 1780. Another engineer, Joshua Wrigley, introduced rotative engines of a similar type four years later at the first cotton mill in Manchester. Pickard had premises in Birmingham, Watt's adopted city, and

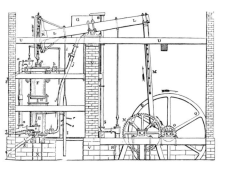

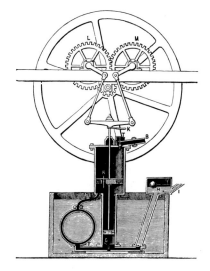

Many engineers beside Watt were working on rotative motion in the late-18th century. In 1758 Keane Fitzgerald devised a combination of toothed wheels, racks and pinions (top left) to be operated by the piston rod and engine beam. Like other similar schemes, it was wholly impractical. In 1780 Matthew Wasborough successfully patented the application of the crank and flywheel to a Newcomen engine (above left). Francis Thompson in 1792 used wheels similar to Watt's sun and planet gear (top), with a complicated arrangement of chains to connect the beam to the piston rod, and Edward Cartwright followed in 1798 with a system of two cranked wheels (above), one of which drove a pinion on the shaft of the fly-wheel while the other helped secure a perfectly rectilinear movement of the piston rod and cross-head.

he incorporated a crank and flywheel in his engine. Watt was forced to think along different lines, and in 1781 introduced what he called 'sun and planet gear'. Three years later he patented 'parallel motion', the invention of which he was most proud, to keep a rigid piston rod moving vertically while attached to the end of an oscillating beam. Thus by 1784 he was able to supply engines capable of 'giving motion to the wheels of mills or other machines'.

The rotative engines were in great demand. For the benefit of new owners, Boulton and Watt issued twenty-four directions for working them, insisting particularly on good maintenance. The first direction was 'everything to be kept as clean as possible', and the nineteenth was to clean the boiler 'at least once a month, but if the water be muddy or scurfy, more frequently, as it will otherwise not only be liable to destruction by burning, but will likewise require more coals: *Two evils to be carefully guarded against.*'

Watt was cautious at first about business prospects for rotative power, fearing 'foolish vaunting' of the new type of engine. 'I hear that there are so many mills resting on powerful streams in the North of England,' he once wrote, 'that the trade must soon be over-done.' Boulton, by contrast, had no doubts. As he wrote in 1781, 'the people in London, Manchester and Birmingham are *steam mill mad*', and a year later he pointed out that 'mills, though trifles in comparison with Cornish engines, present a future that is boundless'. Boulton was right to be optimistic. In 1786 he and Watt acquired a valuable shop window in London when two engines were installed at the new Albion Flour Mill on the Surrey side of the Thames near Blackfriars Bridge. Each engine drove ten sets of mill stones. The Albion Mill was burned down in 1791, but the principle of the rotative engine, crucial to the industrial revolution as we know it, had been securely established.

With the rotative engine, as with the pumping engine, Boulton and Watt faced strong competition in some parts of the country from other manufacturers. In Lancashire, as Musson and Robinson have convincingly demonstrated, substantial numbers of millwrights continued to make saleable Newcomen engines long after Watt engines came into use, and there were pirate Watt engines too. Boulton and Watt never enjoyed a monopoly of engine building in the cotton county; probably they built no more than a third of the engines erected there during the quarter century of their patent.

Watt stated that his mind 'ran upon making Engines *cheap* as well as good', but local manufacturers in Lancashire could produce cheaper and did not demand a premium. 'The gross sum which your engines cost at first startles all the lesser manufacturers here,' Watt Junior told his father frankly in a letter from Manchester in 1790. Two years later Francis Thompson was able to sell five of his newly patented engines to textile manufacturers in Lancashire and Nottinghamshire before the textile slump of 1793–94, in spite of the fact that, as John Farey, the historian put it, 'the consumption of fuel was very considerable'.

For inventive range and scientific skill, however, Watt was unrivalled. Not the least of his innovations was the conception of horse-power. Comparisons between steam-power and horse-power had been made by earlier engineers, even including Savery. Smeaton had said of a famous Cronstadt engine that it would raise 27,300 tons of water in 24 hours to the height of 53 feet, 'which is equal to the power of 400 horses'. Watt began thinking in terms of horse-power in 1782, while working on rotative motion, and within a few years his products were frequently referred to as fourteen-horse engines, twenty-horse engines and so on.

A writer in the *Edinburgh Review* complained in 1809 that since 'what is called the horse's power is so fluctuating and indefinite a matter', it was 'perfectly ridiculous to assume it as a common measure'. But Watt defined one horse-power precisely as 33,000 foot-pounds of work per minute, a figure he arrived at after experimenting with 'strong' dray horses, knowing that it represented about fifty per cent more than the average horse could sustain for a working day. In one respect, however, his measure was awkward, since it took no account of differences in working practice between particular engines of the same size and type; it became known therefore as 'nominal horse-power'.

Real horses continued to figure in the relative cost calculations made by manufacturers pondering on whether or not to introduce rotative steam engines, just as they had figured in the cost calculations of mine owners wondering whether to introduce Newcomen engines into pumping and just as they were to figure in late 19th-century calculations about the purchase of automobiles. The first customers were often more numerate than literate. Thus, when Timothy Harris, a cotton manufacturer of Nottingham, decided to power his mill with a Watt engine instead of horses, he said of one of Watt's agents, who had come to look at his cotton mill, 'which no[w] works by Horses', that while he 'seems to like the siteation of the place vaary whell for to Erect one of your steame Engines' he could not give him any answer about 'The Expense that will attend such an Engine'. 'We now make use of Eight Horsers,' he went on, 'but shold we agment our meacheanery to take 10 Horsers to work the said mill I shold be glad to kno the differance betwene a mill that wourks Eight and that of 10 and the Quantity of Cole it will take to worke the said mill with a day reconing twelve howers for each day.'

Another method Watt used for measuring performance was the steam engine indicator, introduced by him in about 1790 and brilliantly improved in 1796 by J. Southern, one of his assistants. The indicator, which incorporated a little cylinder and piston, measured the varying pressure inside the engine cylinder. 'When the engine was at rest and steam off,' Watt wrote, 'the indicator-piston stood at the same level as when detached from the engine, and the pointer stood at O on the scale. When steam entered, the piston rose and fell with the fluctuations of pressure and a pencil traced a paper record of the movement of the piston.' The device was so useful that it was described as the 'engineer's stethoscope'. 'Indicated horse-power' was to become the mid 19th-century measure.

One of Watt's most important inventions was 'the governor for regulating the speed of the Engine'. This device, which foreshadowed 20th-century automation, consisted of two balls which flew outwards when the speed increased and lifted a sleeve which controlled a butterfly valve in the steam pipe. This was the first regulator in history to be employed extensively, although it was an adaptation of a device used earlier in wind-driven flour mills, and it was to figure with various modifications not only in all future steam engines but in all turbines.

A POWER LOOM

WASHING COTTON CLOTH

MULE-SPINNING

Once the steam engine had been made rotative, it could power all kinds of machinery by belt- and later gear-drive. First applied to the textiles industry, it drove spinning mules (so called because they were a hybrid of earlier

SCOURING WOOLLEN CLOTH

GIG MILL FOR TEAZLING WOOL

inventions), power looms and the many finishing machines, including scouring and washing machines to remove the grease and size used in spinning and weaving, and gig mills for teazling or raising the nap on woollen cloth.

The great 19th-century physicist James Clark-Maxwell, perhaps the first scientist to recognise the importance of 'feedback', a key concept in cybernetics, wrote a paper about the governor, and a 20th-century physiologist claimed that 'Watt had incidentally constructed in principle the first working model of a reflex circuit similar to [that described] in the organisation of sense organs, nerves and muscles'.

Significantly, Watt did not seek to patent either the governor or the steam engine indicator. The governor was allowed to come into general use; the indicator was for years kept secret from all but Boulton and Watt's staff. Not until Farey found members of the firm using one in Russia in 1826 and brought the instrument back with him to England, did its existence become generally known.

The reason why Watt left some of his inventions unpatented was that he disliked the possibility of becoming caught up time and time again in protracted litigation, particularly over such mechanisms as the governor which had pre-existed in an inferior form. He would have preferred the protection of patent law to be offered to everyone who combined together 'old instruments or machines so as to produce new effects, or to make them more extensively useful to the publick', but he knew that such an extension of the law would always be challenged.

There has been far more discussion of the effects of Boulton and Watt's twenty-five year patent right (to which must be added the six years that Watt's own patent had run before 1775) than there has been of Savery's master patent. The discussion began, often acrimoniously, during the term of the patent itself, when in addition to a costly five-year legal battle with Edward Bull, a Cornish infringer of the separate condenser, Watt brought various other cases particularly involving the Hornblower brothers. Such litigation often exasperated him, but it opened up general questions about patents and the public interest which were not always to be decided in Watt's favour.

One voluble contemporary critic of Watt was Joseph Bramah, 'the eminent Water Closet Maker' as Watt described him, who 'called himself an Engineer' (and in fact proved himself an outstanding one). Bramah argued that Watt had taken out his patent rights not for what he had invented, but for what he might invent in the future. Watt was eloquent in his own defence, although he admitted discerningly in 1794 that he was not so 'presumptuous as to think that there were not and are not Numbers of Mechanics in this Nation who from the same, or even fewer, Hints would have compleated a better Engine than he did.'

During the early 1790s he wrote a general statement called 'Thoughts upon Patents, or exclusive Privileges for new Inventions', in which he expounded his central argument. 'Few projectors and . . . few men of ingenuity make fortunes, or even çan keep themselves on a footing with the tradesman who follows the common tricks,' Watt began. The 'man of ingenuity' had to 'devote the whole powers of his mind to one object . . . and had to persevere in spite of the many fruitless experiments he makes. . . . It is argued by the opponents of patents that, they tend to cramp ingenuity, by circumscribing the artist to the use of the arts which prescription has made public property. Those who argue in this manner have too narrow notions of the human mind, and of the objects on which it can exert itself.'

Watt denied that his patent had impeded progress, arguing that the 'improvements which have been made within the last 50 years *surpass all which have been done in an equal period of time*'. Of the Cornish mine owners, he wrote on another occasion: 'They charge us with establishing a monopoly, but if it is a monopoly, it is one by means of which their mines are made more productive than they ever were before'. Confident in his case, Watt even attempted to extend his patent still further after it expired in 1800. Watt's arguments have cut little ice with recent historians. His distinguished biographer and admirer, Dickinson, himself came to the conclusion that the twenty-five-year patent was 'unduly long in the public interest' and 'tied down progress to Watt's chariot'. He had in mind, in part at least, Watt's unwillingness to pursue experiments with high-pressure steam. Professor T. S. Ashton in a brilliant essay on the industrial revolution considered that Watt's opposition to high-pressure steam may have held back the age of railways by a generation: 'the authority he wielded was such as to clog engineering enterprise.'

More steam engines of every kind were built between 1790 and 1800 than in the rest of the 18th century put together. Twelve English counties obtained their first steam engine during this decade, and by the beginning of the new century only Bedford, Buckingham, Hereford, Huntingdon, Rutland and Sussex were without an engine.

Yet with all this expansion of engine building, scientists were slow to turn from the kind of practical questions which engineers considered to general questions of heat and work. In the very last year of the old century Count Rumford, born Benjamin Thompson in Massachusetts, founded in England the Royal Institution, which aimed to popularize science. Rumford studied the heat of friction, but the laws of thermodynamics were worked out not in the 18th but in the 19th century, and not in England but in France and Germany.

The key figure in early 19th-century scientific work was the Frenchman Sadi Carnot who published in 1824 his important *Reflections on the Motive Power of Fire and on Machines fitted to develop that Fire*. Carnot's main contribution to the scientific knowledge of heat was a cycle of heat and work exchanges in an ideal steam engine – the so-called Carnot cycle. The laws of thermodynamics were to be worked out on the basis of mathematical theory developed by Carnot. One major English contributor to this work was J. P. Joule in Manchester, who during the 1840s measured the mechanical equivalent of heat.

By the time that Carnot produced his study, a new non-technical concept was beginning to take shape: the 'industrial revolution', a catch-all term which became a label for a number of related economic and social trends. Chief among these trends were the substitution of machines for human effort and skill; the substitution of inanimate for animate sources of power; and the exploitation and transportation of raw materials and manufactured products.

The term 'industrial revolution' was not used until 1827 – by another Frenchman, Adolphe Blanqui – and it was not popularized until much later in the 19th century. Yet soon after the French Revolution, which began in 1789, contemporaries were beginning to compare the political changes in France with the technological and social changes taking place in England. Both seemed to have revolutionary implications. Britain was able at first to take advantage of the troubles in

CUTTING PAPER

PAPER-MAKING

FLAKE COCOA MILLS

Manufacturing processes ranging from paper-making and metal-working to confectionery and printing were soon harnessed to steam. In the continuous-process paper-making machine above, a vat at the right-hand end contains pulp which is fed on to an endless web of wire cloth, allowing water to drain off. The pulp then passes through steam-filled cylinders for drying

ENVELOPE-FOLDING MACHINE

NEEDLE-POINTER AT WORK

and is finally wound on to a roll. Cutting and folding machines turn it into envelopes. The steam engine was also well suited to driving cocoa mills in which powdered cocoa was turned into flakes, and to powering grindstones which would sharpen a handful of 50 to 100 needles in 30 seconds (creating a noxious dust).

France and of the wars which followed to develop its industrial potential, though eventually the two revolutions were seen as converging rather than contrasting, technological progress providing a strong material base for the democratic society which the French Revolution proclaimed.

It is important to reiterate that steam power by itself did not cause the industrial revolution. As Eric Roll put it well in 1930 in a study of the business operations of Boulton and Watt: 'The question as to whether steam power alone was responsible for the industrial revolution was debated even before the series of profound economic changes of the last half of the 18th century came to be known by that term. It is now generally recognised that the factory system, together with all the forces that moulded the modern industrial age, originated before and apart from the application of steam power to industry. At the same time it is admitted that without the existence of some such agent neither the rapidity nor the completeness of the development of the modern economic system can be explained.'

More recently, quantitative economic historians in Britain, following in the wake of academic colleagues in the United States, have tried to get behind such tidy generalization. S. W. von Tunzelmann has entered into an 'as-if' world, trying to measure what would have happened had Watt not invented his steam engine. His own conclusion, based on elaborate mathematical calculation, is that dependence on earlier or improved forms of the Newcomen engine, plus a combination of old animate power sources, would have produced a rate of industrialization almost as fast as that which actually took place. The decisive breakthrough in his view was not the application of the Watt engine.

Such an 'as-if' society was not, of course, the industrial society our ancestors came to know. For them steam engines were a source of increased economic strength and flexibility. It was the great merit of steam power, they felt, that it could be increased at will. Without steam other changes central to the course of the industrial revolution could not have happened. Steam changed the location of industry, encouraging a move from the banks of rivers to the expanding towns. It substituted concentration for dispersion, with all the social consequences which that entailed. Above all, it reinforced a new sense of system in the industrial order, starting with the management of the factory itself. 'We are systematising the business of engine making,' Matthew Boulton wrote in 1778, 'as we have done before in the button manufactory.'

One of the first 20th-century academic writers on the industrial revolution, another Frenchman, Pierre Mantoux, emphasised in the 1920s how steam gave unity to the modern factory system. Before steam 'separate industries were much less interdependent than they are now. . . . The use of a common form of motive power, and especially of an artificial one, thenceforth imposed general laws upon the development of all industries.' Yet even the word system fails to do justice to the dynamic impulse that flowed through society after the introduction of steam, influencing both new industrialists and their workers. If the breakthrough may have been exaggerated in strictly economic terms, in other ways it cannot be over-emphasized. Steam marked a psychological break with the past. It propelled a gospel which, like Boulton and Watt's steam engines, was to be carried round the world.

'By-and-bye a Man will go a hunting after breakfast upon his Tay-kettle,' reads the caption to this print of August 1829. Entitled 'Pat's Comment on Steam Engines', it plays on the theme of the Irish joke as well as poking fun at the Rainhill Trials, held that April to establish whether a steam locomotive could provide power for the new Liverpool and Manchester Railway.

THE INVENTIVE SPIRIT

A steam engine that could move by its own power was a revolutionary innovation in the early 19th century. Inventors responded with a plethora of ideas for transport by land, sea and later by air. Many proposals seemed then, as they do now, to be pure fantasy – and satirists were quick to exploit the vein. But some of the most outlandish contraptions were in fact conceived perfectly seriously.

64/

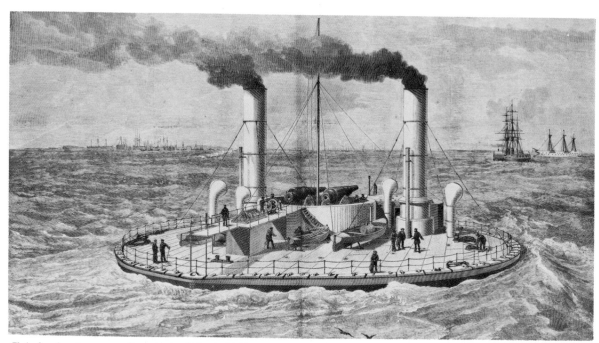

Twin funnels and a gun mounting, with the ship's wheel behind, complete the superstructure of a Russian circular ironclad launched in the early 1870s.

A self-levelling saloon was the centrepiece of this steamer invented in 1874 by the metallurgist Sir Henry Bessemer. It did not live up to expectations.

Giant double-boilered locomotives haul a ship along an inter-oceanic railway at the Isthmus of Panama, proposed in the Scientific American *in 1884.*

'I say Fellow, give my Buggy a charge of Coke,' says a steam motorist, 'your Charcoal is so D-d dear.'

This proposed steam engine for American street railways was disguised so as not to frighten horses.

A Delight in Novelty

The age of motoring, complete with different grades of fuel, was amusingly previewed by H. Alken in a print (left) from his series the 'Illustration of Modern Prophecy'. Practical developments, while slower, were sometimes no less amusing. Several inventors, believing that powered road wheels would slip, designed a kind of vehicle (right) propelled by jointed legs and feet.

This leg-driven carriage was patented in 1824 by David Gordon, who carried out trials of the system.

Clouds of smoke rise from the frozen Hudson River as steam saws invented by Chauncy A. Sager, an American engineer, cut ice to be stored for summer use.

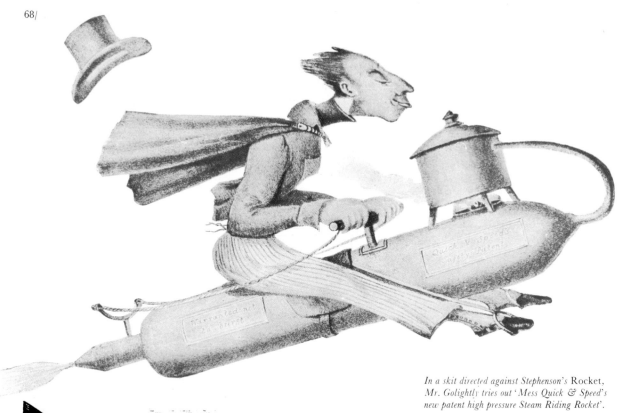

In a skit directed against Stephenson's Rocket,
*Mr. Golightly tries out 'Mess Quick & Speed's
new patent high pressure Steam Riding Rocket'.*

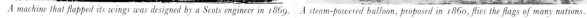

A machine that flapped its wings was designed by a Scots engineer in 1869.   *A steam-powered balloon, proposed in 1860, flies the flags of many nations.*

This was one of many designs for steam-driven dirigibles produced during the 19th century. The few airships that were built failed through lack of power.

CHAPTER THREE

THE GOSPEL OF STEAM

And now, great masters of the realms of shade,
To end the task which called us down from air,
We shall present, in pictured show arranged,
Of this your modern world, the triumphs rare,
That Gryllus's benighted spirit
May wake to your transcendent merit,
And, with profoundest admiration thrilled,
He may with willing mind assume his place
In your steam-nursed, steam-borne, steam-killed,
And gas-enlightened race.

THOMAS LOVE PEACOCK
Gryll Grange (1860)

Smoke rising from the factory chimneys of Sheffield in the 1930s creates an archetypal industrial scene. Air pollution had been singled out for attack by the critics of steam power as early as the 18th century. The many apostles of steam, including Watt himself, could find no solution to the problem.

During the last decades of his life a lonely James Watt wrote that since he knew that all men must die, he submitted himself to 'the decrees of Nature . . . with due reverence to the Disposer of events'. It was not the kind of remark he would have made as a young man. Indeed, he is reported to have said once that 'Nature can be conquered if we can but find her weak side'. Nor did his contemporaries see him as a great respecter of nature. Most of the tributes paid to him after his death in 1819 emphasized how he had transformed man's sense of nature by converting, adapting and applying it for the use and convenience of man. Contemporaries argued too, that through his invention and enterprise Watt had acted himself as a disposer of events, for by 1819 not only the rhythm of production but the whole outlook of society was being changed by the impact of steam-based industrialization.

At that time by no means the whole of British industry – even of the textiles industry – was operated by steam power; the use of steam for locomotion was still in its early stages; and in many great English industrial centres such as Birmingham, where Watt lived and worked, the application of steam was slow in the large number of small workshops that specialized in the metal trades. The general diffusion of steam power (described in Chapter 5) was still in the future. Yet already writers of both prose and poetry were propagating a 'gospel of steam', enthusiastically portraying the potential of the new technology to transform almost every branch of human activity.

Watt's death provided a suitable occasion for the first public airing of the gospel. He himself was identified in both classical and romantic terms as the hero of steam. The young Whig leader, Henry Brougham, later Lord Chancellor, wrote in his epitaph for Watt's statue in Westminster Abbey: 'Not to perpetuate a Name, which must endure while the Peaceful Arts Flourish, but to show that Mankind have learned to honour those who best deserve their gratitude, the King, his Ministers, and many of the Nobles and Commons of the Realm, raised this Monument to JAMES WATT, who directing the Force of an Original Genius, early exercised in Philosophic Research to the Improvement of the Steam Engine, enlarged the Resources of his Country, increased the Power of Man, and rose to Eminent Place among the most illustrious Followers of Science and real Benefactors of the World.'

For once, Tories as well as Whigs – with Sir Robert Peel, son of a Lancashire calico manufacturer, prominent among them – were in agreement with Brougham's tribute. More surprisingly, perhaps, the romantic poet William Wordsworth, whose compelling personal vision of the inherent strength of nature was very different from Watt's willingness to seek out nature's weak side, described Watt as 'perhaps the most extraordinary man that the country has ever produced' and emphasised both the 'magnitude' and the 'universality' of his achievement. Likewise, Thomas Carlyle, prophet of the age, pictured a heroic Watt in striking language: 'this man with blackened fingers, with grim brow . . . searching out, in his workshop, the Fire-secret.'

An earlier writer, Erasmus Darwin, whose elegantly turned verses became unfashionable at the time when Wordsworth's simple poetic style came into vogue, had produced some of the best known verse about the potential of steam power in his *The Economy of Vegetation* (1792):

Stressing the useful powers of steam, a Leeds manufacturer explained in the caption to this catalogue illustration that he had 'sought to meet every requirement by constructing his Machines in various forms and variety, for the breaking and crushing of every description of hard material to any required size'.

'Nymphs! You erewhile on simmering cauldrons play'd,
And call'd delighted SAVERY to your aid:
Bade round the youth explosive STEAM aspire,
In gathering clouds and wing'd the wave with fire:
Bade with cold streams the quick expansion stop,
And sunk the immense of vapour to a drop –
Press'd by the ponderous air the Piston falls
Resistless, sliding through its iron walls;
Quick moves the balanced beam, of giant birth,
Wields his large limbs, and nodding shakes the earth.
The Giant-Power from earth's remotest caves
Lifts with stronger arms her dark reluctant waves:
Each cavern'd rock, and hidden den explores,
Drags her dark coals, and digs her shining ores . . .
Soon shall thy arm, UNCONQUER'D STEAM! afar
Drag the slow barge, or drive the rapid car:
Or on wide-waving wings expanded bear
The flying-chariot through the fields of air.'

In a prose note Darwin added that 'there seems no probable method of flying conveniently but by the power of steam or some other explosive material, which another half century may probably discover' and that thanks to Boulton and Watt, both of whom he knew well, steam was already being applied to 'almost every purpose'.

What came to be thought of in the 19th and 20th centuries as 'the two cultures' were spanned in writing like this, as they were by Wordsworth, who wrote in 1800 of enlisting 'Imagination under the banner of Science' and prophesied that 'if the labours of men of science should ever create any material revolution, direct or indirect, in our condition . . . the Poet will sleep no more than at present: he will be ready to follow the steps of the Man of Science.' There was a further visual link. Beauty could be associated with machines as well as with mountains, so inspiring to the Romantic sensibility. 'There cannot be a more beautiful and striking exemplification of the union of science and art than is exhibited in the steam engine,' Benjamin Heywood, the Chairman of the Manchester Mechanics Institute, told the members in his opening address in 1825, adding more controversially that 'there is no art which does not depend more or less on scientific principles'.

Such talk was not short-lived. More than twenty years later, when steam engines were to be found all over the world, E. M. Bataille, the French author of a *Traité des Machines à Vapeur*, went even further and claimed enthusiastically that 'the part which Watt played in the mechanical application of the force of steam can only be compared to that of Newton in astronomy, and of Shakespeare in poetry'. And, he went on to ask, 'is not invention the poetry of science?'

A positive answer to this question, supplied with confidence, provided the first article in the gospel of steam. Belief in the creative power of invention was proclaimed in many different quarters. 'Invention, physical progress, discovery are the war cries of today,' wrote the *People's Journal* in 1847. 'Prospero can send his Fire-demons panting across all oceans,' exclaimed Carlyle, and Henri Rollin, in his *Introduc-*

A Russian edition was among the foreign-language printings of the 'Descriptive Catalogue of Agricultural Machinery Manufactured & Sold by Ransomes, Sims & Head' of Ipswich. It was published in 1879. Five years earlier two of the firm's partners had demonstrated a straw-burning portable engine before Tsar Alexander II.

74

tion to the Arts and Sciences of the Ancients, added a note of rhetoric: 'Of what utility to us at this day, are Nimrod, Cyrus, Alexander or their successors, who have astonished mankind from time to time?' he asked. 'With all their magnificence and vast designs, they are returned into nothing with regard to us. They are dispersed like vapours, and have vanished like phantoms. But the INVENTORS of the ARTS and SCIENCES laboured FOR ALL AGES. We enjoy the fruits of their application and industry – they have preserved for us *all the conveniences of life* – they have converted all nature to our uses.'

Such pride in invention raised expectations and stimulated prediction about the future. Wordsworth's sonnet 'Steamboats, Viaducts and Railways' (1833) pointed to the 'crown of hope' proffered by the new works of engineering:

'Nor shall your presence, howso'er it mar
The loveliness of Nature, prove a bar
To the Mind's gaining that prophetic scope
Of future change.'

Likewise, *Les Voyageurs* (1844) by Victor Hugo told poets how iron and steam had changed the world: the 19th century was *'grand et fort'*, and the appeal of the picturesque was giving way to the appeal of the dynamic. Like Wordsworth, Hugo did not believe that spiritual power and material power necessarily went together, yet then and later he spoke so enthusiastically about steam that he did much to spread the gospel.

An English provincial poet, Ebenezer Elliott, the Corn-Law Rhymer, whose poems attacking the Corn Laws impressed Carlyle and stirred radical crowds in England's great industrial cities, made the most of the potential of the 'useful powers' of steam. In one of his poems an imaginary blind character, Andrew Turner, is told of the miracles of steam at Sheffield:

'Come blind old Andrew Turner! link in mine
Thy time-tried arm, and cross the town with me;
For there are wonders mightier far than thine;
Watt! and his million-feeding enginery!
Steam – miracles of demi-deity!
Thou can'st not see, unnumber'd chimneys o'er
From chimneys tall the smoky cloud aspire;
But thou can'st hear the unwearied crash and roar
Of iron powers, that urged by restless fire,
Toil ceaseless, day and night, yet never tire
Or say to greedy men, "Thou dost amiss".
Oh, there is glorious harmony in this
Tempestuous music of the giant, Steam,
Commingling growl, and roar, and stamp, and hiss,
With flame and darkness! Like a Cyclop's dream....

Enthusiasm for the endless vistas of human transformation opened up by steam led on naturally to the second article in the gospel: the universality of the 'new powers'. As had long been forecast, the influence of steam was expected to be global, not national. Elliott in his poem turned easily next to this theme, introducing blind Andrew to horizons stretching far beyond Sheffield:

A 366-foot octagonal chimney shaft, erected at Bolton in 1842, contrasts with more elaborate designs for both chimneys and ventilating shafts by the architect Rawlinson. Though utilitarian in purpose, chimneys were the totem poles of the firms that built them, and Classical, Gothic and Norman architectural styles were ransacked to provide fitting adornments – evidence of the once fertile union between technology and the traditional arts.

'And rolling wide his sightless eyes, he stands
Before this metal god, that yet shall chase
The tyrant idols of remotest lands,
Create science in the desert, and efface
The barren curse from every pathless place.'

Steam was to penetrate the unexplored and undeveloped world as well. Elliott called another of his poems 'Steam in the Desert':

'Engine of Watt! unrivall'd is thy sway.
Compared with thine, what is the tyrant's power?
His might destroys, while thine creates and saves
Thy triumphs live and grow, like fruit and flower.'

All mankind would benefit from steam. By the time Watt died, it was known and acclaimed that the making and improvement of steam engines was already an international activity; and in a French eulogy delivered before the Academy of Sciences on the occasion of the twentieth anniversary of Watt's death, the well-known astronomer Dominique François Arago selected Watt as the first hero of international progress. 'Let us reckon upon the future! A time will come, when the science of destruction shall bend before the arts of peace. . . . Then Watt will appear before the grand jury [of mankind]. . . . Every one will behold him, with the help of his steam-engine, penetrating in a few weeks into the bowels of the earth . . . excavating vast mines, clearing them in a few minutes of the immense volumes of water which daily inundated

Workers employ a steam navvy to construct a sewer in Shoreditch, London, in 1862. Equipment of this kind fulfilled the prophecy that steam would lighten the drudgery of human toil.

The commitment of labour to technical progress and the useful arts is well depicted in this membership certificate of the Steam Engine Makers Society.

The chimneys of *Sharp & Brown, wire manufacturers, tower above Birmingham (right). Early-19th-century factories such as this, served by canal barges and horse-drawn traffic, swelled the urban industrial labour force and created health dangers that were well documented in official and unofficial reports. The employment of factory children (above) broke up family life and led to agitation for age restrictions and shorter working hours.*

SHARP & BROWN WIRE

…UFACTURERS FAZELEY STREET BIRMINGHAM.

One of the lighter attractions of the steam age, this fairground organ built by Gavioli of Paris was originally driven by a showman's traction engine.

them. . . . The population, well supplied with food, with clothing and with fuel, will rapidly increase. . . . In a few years hamlets will become great towns; in a few years boroughs, such as Birmingham . . . will take their place among the largest, handsomest and the richest cities of a mighty kingdom.'

The President of Anderson's College in Glasgow had been more daring when he addressed a meeting held in 'Watt's city' in 1825 to discuss a proposal to erect a monument to Watt. Echoing Erasmus Darwin, he exclaimed to the applause of his audience that 'the time is not far distant when chariots winged with fire will be seen flying over metallic pavements through all the populous districts of the empire' – and they would unite not just the Empire, but Mankind.

Later in the 19th century the same sense of universality – and awe – was apparent in the poetry of Rudyard Kipling. By then prediction had given way to fact and steam had actually linked the continents. 'McAndrew's Hymn' (1893) begins:

> 'Lord, Thou has made this world below the shadow of a dream,
> An', taught by time, I tak' it so – exceptin' always Steam.
> From coupler-flange to spindle-guide I see Thy Hand, O God –
> Predestination in the stride o' you connectin' rod.
> John Calvin might ha' forged the same – enormous, certain, slow
> Ay, wrought it in the furnace-flame, *my* "Institutio".'

The fascinating reference to Calvin carries the full flavour of a gospel.

Kipling believed fervently also in the third article of the gospel, which stressed the advantages of mechanization. Steam put an end to age-old drudgery: it took over. This theme was best stated by Andrew Ure, the self-appointed spokesman of the factory system. What had previously been laborious now became 'automatic', he proclaimed, as 'elemental powers' were made 'to animate millions of complex organs, infusing into forms of wood, iron and brass an intelligent agency'. Muscular fatigue was reduced, while 'handicraft caprice' disappeared 'under the safeguard of automatic mechanism'. Within the factory 'the benignant

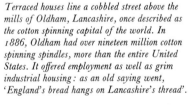

Terraced houses line a cobbled street above the mills of Oldham, Lancashire, once described as the cotton spinning capital of the world. In 1886, Oldham had over nineteen million cotton spinning spindles, more than the entire United States. It offered employment as well as grim industrial housing: as an old saying went, 'England's bread hangs on Lancashire's thread'.

Steam on Military Service

Troops with a steam sapper at Chatham, 1873.

The prophets of steam had said that it would bring peace. In fact, designers adapted the new power to military use even when it was inappropriate. The Garrett steam submarine, invented by a vicar, ran on storage boilers. Steam fed in before a mission gave it a maxi-mum range of twenty to thirty miles. It never saw active service. Nor did the steam tank, which reached only prototype stage. But traction engines, wagons and loco-motives provided haulage, and steam powered the rival navies, in both 19th- and 20th-century wars.

An armoured road train built by John Fowler & Co., Leeds, goes on show before delivery to the War Department in 1900.

A Fowler steam wagon, in field-hospital service in France during the First World War, carries two big steam cylinders for sterilizing infested clothing.

Personnel examine a US steam tank in 1917.

The torpedo boat Hornet *was the fastest vessel in the world in 1893, with a top speed of 27.3 knots.*

A light engine comes ashore in Madagascar, 1942.

Pioneers stand by the open conning tower of G. W. Garrett's submarine Resurgam, *built in 1879.*

A Burrel traction engine from Aldershot smokes on the near bank while another hauls an Army road train across a river in South Africa during the Boer War.

power of steam summons around him his myriads of willing menials, and assigns to each the regulated task, substituting for painful muscular effort on their part, the energies of his own gigantic arm [the same metaphor as Darwin's], and demanding in return only attention and dexterity to correct such little aberrations as casually occur in his workmanship. The gentle docility of this moving force qualifies it for impelling . . . tiny bobbins . . . with a precision and speed inimitable by the most dexterous hands.' In Kipling's poem 'The Ship that Found Herself', steam took over completely: it did not even need 'dexterous hands'.

The fourth article of the gospel was that steam brought rising living standards and economic growth. In an age of steam power 'the barest necessities of life' could easily be supplemented by what had hitherto been conceived of as luxuries. 'Steam engines,' Ure argued – and he was propounding conventional wisdom – 'furnish the means not only of their support but of their multiplication. They create a vast demand for fuel; and, while they lend their powerful arms to drain the pits and to raise the coals, they call into employment multitudes of minders, engineers, ship-builders, and sailors, and cause the construction of canals and railways; and, while they enable these rich fields of industry to be cultivated to the utmost, they leave thousands of fine arable fields free for the production of food for man, which must have been otherwise allotted to the food of horses.' Much though Marx disliked Ure's doctrines, he agreed with him on these points. The *Communist Manifesto* of 1848, written with his friend Friedrich Engels who lived in smoky Manchester, is particularly eloquent in its celebration of the 'subjection of nature's forces to man's machinery' and speaks of 'whole populations conjured out of the ground'. 'What earlier century,' its authors asked, 'had even a presentiment that such productive forces slumbered in the lap of social labour?'

The gospel of steam had a link with another gospel also: that of free trade. As Ure explained, 'Steam engines, by the cheapness and steadiness of their action, fabricate cheap goods, and procure in their exchange a liberal supply of the necessaries and comforts of life, produced in foreign lands.' What technology could achieve, men should not frustrate by interposing artificial barriers, particularly if they were designed to protect traditional aristocrats against enterprising mill-owners.

Free trade would bring peace. 'The hour is coming, hastening with the momentum of ages,' wrote M. A. Garvey in his book *The Silent Revolution; or the Future Effects of Steam and Electricity upon the Condition of Mankind* (1852), when 'the argument of murder, rapine, famine and pestilence shall be banished from amongst men; consigned to the chamber of horrors, in which history preserves the memorial of crime.'

The fifth article in the gospel of steam, that steam opened up new horizons for the working class, could be related either to the Marxist analysis that steam power fused working-class consciousness, or to the individualist analysis behind yet another gospel: self-help. Sometimes both types of analysis overlapped. Thus in the preface to *A Descriptive and Historical Account of Hydraulic and other Machines for Raising Water . . . including the Progressive Development of the Steam Engine*, a remarkable book published in 1852 and illustrated by nearly three hundred engravings,

Fun of the Fair

From the 1870s until after the First World War, fairgrounds were enlivened by the smell of coal smoke, the blare of organs and the sheer exhilaration of steam-powered rides on scenic railways, gondolas, dragons and gallopers.

Many rides had centre engines with chimneys projecting from the big-top awnings. Others were powered electrically by showman's engines with dynamos and extended chimneys to keep smoke off the customers and the varnished paintwork of the rides.

John Evans's live-rail dragon scenic (bottom right) was typically elaborate. It used two engines, one to drive the cars (each weighing two tons and carrying 12 people), the other to power the organ, the lights and the centrifugal pump that kept water flowing down the steps of a painted waterfall decorated with real ferns.

At the 1911 Nottingham Goose Fair, the market

Harry Gray's steam yachts, powered by a Savage's engine and each carrying 30 fairgoers, draw the crowds at Hampstead Heath, London, in the 1920s.

Drawn by a traction engine, a calliope belching black smoke arrives at a steam rally in Milwaukee.

place is filled with steam and smoke from scenic rides.

An engine with crane gets up steam to dismantle John Evans's scenic dragons at Newcastle in 1922.

the American Thomas Ewbank had as much to say about mechanics as about inventors, and his tone was different from Ure's. 'Few classes have a more honourable career before them than intelligent mechanics,' Ewbank asserted. 'Under the old regime' (before the industrial revolution and the French revolution) operators had been treated as 'virtual serfs'. In future, they would 'exert an influence in society commensurate with their contributions to its welfare'. Science and the arts were 'renovating the constitution of society'.

A writer in the *Westminster Review* in 1854, praising 'the diligence, the dexterity and the ingenuity of English workmen', went so far as to claim that they had been the main source of invention. 'Deduct all that men of the humbler classes have done for England by way of inventions only, and see where she would have been but for them.' Inventions were the result of 'much patient plodding and persevering ingenuity on the part of our mechanics'. With steam power in mind, the writer quoted an authority who had claimed that by art, science and mechanical skill the 'useful productions' of six counties in England – Lancashire, Yorkshire, Cheshire, Staffordshire, Nottinghamshire and Leicestershire – by then greatly exceeded what could have been effected earlier by 'the entire human family by means of physical labour alone'.

Andrew Ure himself argued that 'occupations which are assisted by steam-engines require for the most part a higher, or at least a steadier species of labour, than those which are not; the exercise of the mind being then partially substituted for that of the muscles, constituting skilled labour, which is always paid more highly than unskilled'. Yet, like other defenders of the new industrial order, Ure was concerned that operatives either did not necessarily see the matter in this light or that through 'combining' and organising strikes they introduced friction into what by its nature was a beautifully smooth system. Thomas Carlyle, by contrast, sympathised with those workers who were not able to develop the skills of steam power operatives. 'It is consistent that the wages of "skilled labour", as it is called, should in many cases be higher than they ever were: the giant Steamengine in a giant English nation will here create violent demand for labour, and will there annihilate demand. But, alas, the great portion of labour is not skilled: the millions are and must be skill-less . . . hewers of wood and drawers of water, menials of the Steamengine.'

Carlyle was excited by 'the clank of innumerable steam engines', but at the same time fearful. 'The whole is not without its attractions, as well as repulsions,' he wrote, although in his *Signs of the Times* he objected strongly to people themselves being treated as machines.

The makers of the gospel of steam knew that like all gospels it had not won authority everywhere and at once. Just as the gospel of free trade was a militant gospel because of the strength of protectionist interests and attitudes (and the gospel of self-help could be equally militant because of the prevalence of 'idleness', aristocratic or proletarian), so the gospel of steam derived power from the knowledge of its disciples that ranged against it were sceptics and hostile unbelievers.

The sense among advocates of steam that they were propounding a new gospel encouraged vigorous efforts to discredit old beliefs and attitudes. Rollin's *Hymn to Invention*, in which he selected inventors as

Another fair over, a Burrell showman's engine named Challenger *stops on a Lincolnshire road to take water. The long convoy of trailers behind carries a bioscope show, a form of travelling cinema common in the early 1900s. Power for the projector was provided by the dynamo, covered with a tarpaulin, on the front of the engine.*

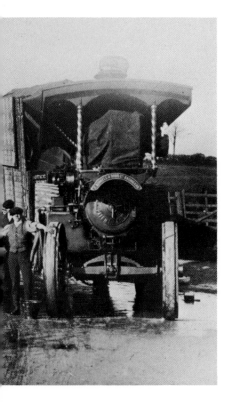

benefactors of mankind, ended: 'Yet all our admiration turns generally on the side of those heroes in blood, while we scarce take any notice of what we owe to the INVENTORS OF THE ARTS'. Thomas Ewbank chose as the motto of his book a passage from the 18th-century Scots historian, the Presbyterian Minister, William Robertson: 'It is a cruel mortification in searching for what is instructive in the history of past times, to find that the exploits of conquerors who have desolated the earth, and the freaks of tyrants who have rendered nations unhappy, are recorded with minute and often disgusting accuracy – while the discovery of useful arts, and the progress of the most beneficial branches of commerce are passed over in silence, and suffered to sink into oblivion.'

Ewbank was the kind of writer, who, like the Rev. Dionysius Lardner, was re-arranging history in the light of his new gospel. So, too, supremely, was Samuel Smiles, whose five-volume *Lives of the Engineers*, the fourth volume of which dealt with Boulton and Watt, took his eager readers, as one reviewer remarked with some exaggeration, on 'a most interesting, because untrodden, walk in literature'. Smiles's 'clear manly style' was praised. So too was his skill in sifting 'a vast mass of authentic documents'. For the *Birmingham Daily Gazette* the Boulton and Watt volume was 'a worthy monument'. 'The lives of both men, so variously but so richly endowed – the work which they did, and the courage with which they met and overcame enormous difficulties – are more exciting than the story of a great battle. In fact, the two lives might be called the story of a great battle – the contest of genius and courage with the forces of nature, in which the former obtained the victory.'

Smiles had many working-class readers, and his volumes on the engineers were often distributed as prizes, as were his *Self Help* (a best-seller), *Thrift*, *Character* and *Duty*. Mechanics' institutes and Sunday schools were very happy to distribute them to the young. There was, indeed, a whole literature about steam designed for the young, including the very young:

' "Only fancy, Aunt Helen," added Charles, "that Uncle Harry was in Paris yesterday and will be at home today. Is it not wonderful? What did men do before there were railroads and steam boats?"

"Went by coach or van, or in their own carriages, or on foot as the case may be."

"What a waste of time," exclaimed Charles. "It was a good thing that steam was discovered." ' This passage from a book called *The Triumphs of Steam*, written by an anonymous author (not Frith), was published in 1859, the same year as Smiles's *Self-Help*.

The attitudes and achievements which Smiles held up as an example were very much the same as those praised by Ebenezer Elliott – indeed, the two men worked closely together during the radical movement of the 1840s. It was during that stormy decade – a decade not so much of blue prints for the future but of blue books, official reports on pressing economic and social issues – that the gospel of steam was challenged most bitterly by writers, by factory operatives and by the large but declining force of handloom weavers who had been displaced by the machine. Two aspects of steam were spotlighted for attack: its impact on the environment and on human relations.

Complaints about smoke had preceded the development of the steam engine: they were heard in London in the 1660s, for example, when the diarist John Evelyn wrote his *Fumifigium* and William Petty, the statistician, grumbled about 'the fumes, steams and sinks' of the east of the city. There was a brilliant attack in 1725 on the Savery engine installed in the York Buildings in London. The noise of the monster, it was said, would reach Calais and 'being of a *huffing, snuffing* temper' it would 'dart out of its nostrils perpendicularly up to the skies two such vast, dense and opake columns of smoke, that those who live in the Borough will hardly see the sun at noonday'. Later James Watt spent time trying to invent a patent smokeless fireplace and corresponded with Dr. Thomas Perceval, an enlightened and conscientious Manchester doctor who was concerned about the effects of pollution in Manchester, but nothing came of these efforts. By the 1840s and 1850s, although concern had not disappeared, neither had the smoke. In Charles Dickens's 'Coketown', magnificently sketched in his novel *Hard Times* (1854), the air was free from noise and relatively free from smoke only on a Sunday. Even then a black mist persisted over Coketown, but 'engines at pits' mouths and lean old horses that had worn the circle of their daily labour into the ground were alike quiet: wheels had ceased for a short space to turn; and the great wheel of earth seemed to revolve without the shocks and noises of another time'.

For Dickens and for other writers, including John Ruskin and William Morris, questions of human environment and of human relations were connected. 'The piston of the steam engine worked monstrously up and down like the head of an elephant in a state of melancholy madness,' Dickens wrote. Hot or cold, wet or fine, the working day depended for its rhythms not on people but on 'all of the melancholy, mad elephants polished and oiled up for the day's monotony'. The owners of the engines could calculate 'to a single pound weight' what the steam engine would do, but were helpless in front of 'the unfathomable mystery' of each of their workers, 'rating 'em', as one worker complained, 'as so much Power, and reg'lating them as if they was figures in a soom [sum] or machines'. Nature might be harnessed, but human nature could not be. As Léon Faucher, a French observer of Manchester put it in 1844, 'Amid the fogs which exhale from this marshy district and the clouds of smoke vomited forth from its numberless chimneys, labour presents a mysterious activity, somewhat akin to the subterranean action of a volcano.'

John Ruskin objected both to the gloom of the factory districts and 'the depressing and monotonous circumstances of English manufacturing . . . Blanched Sun – blighted grass – blinded man'; and William Morris lived long enough through a century of technical and social change to go well beyond Ruskin and preach socialism. His vision had much in common with that of the young Marx long before he had read any Marx, and during the 1880s he was drawn into socialist politics. It was before he became a socialist, however, that he wrote at the beginning of his much-read poem *The Earthly Paradise*:

'Forget six counties overhung with smoke,
Forget the snorting steam and piston smoke,
Forget the spreading of the hideous town:
Think rather of the pack-horse on the down,
And dream of London, small and white and clean.'

In 1869, its fourth year of publication, the English Mechanic featured on its cover a steam velocipede which could, if the artist was to be believed, take a couple on a pleasant ride into the country. The magazine set out to make knowledge of technology available to the millions.

Morris objected both to steam technology and to its ownership in private hands, and his gospel of art and politics won disciples in his own time and in the late 20th century.

The attack on steam by factory operatives was almost as old as the attack on smoke. Many operatives complained long before the 1840s that they could no longer pace their own work, but had become tied to the rhythms of steam. By the end of the 18th century some of them were chafing against the rules, regulations and fines. A number physically destroyed looms, whether powered by steam or by water.

Shock dominated the attitudes of the first generation of workers exposed to the new steam technology. This reaction is still evident in a poem by E. P. Mead called 'The Steam King' which was quoted by Engels in his *Condition of the Working Classes in England* (1845). It began:

'There is a King, and a ruthless King;
Not a King of the poet's dream:
But a tyrant fell, white slaves know well,
And the ruthless King is steam.'

Taking up the image of the arm, which had been used by so many earlier writers, Mead went on:

'He hath an arm, an iron arm,
And tho' he hath but one,
In that mighty arm there is a charm
That millions hath undone.
Like the ancient Moloch grim, his sire
In Himman's vale that stood,
His bowels are of living fire,
And children are his food.'

Before demanding the end of King Steam's reign, Mead condemns his 'satraps':

'His priesthood are a hungry band,
Blood-thirsty, proud and bold;
'Tis they direct his giant hand
In turning blood to gold.'

If the introduction of steam power into industry was a protracted, often slow process, nonetheless it generated complaints during the first half of the 19th century on the grounds that it was disturbing ways of life far too abruptly. Thus, Edward Tufnell in an influential pamphlet of 1834 complained that 'the forced and premature adoption' of steam power led to the displacement of labour with 'inconvenient rapidity'. 'Instead of improvement proceeding by those gently varying gradations which characterize its natural progress, it advances, as it were, *per saltum*, and comes upon the workman unprepared for the change which his course of life must subsequently undergo.' On very similar lines a reviewer of a number of books stated that 'machinery, like the rain from heaven, is a present blessing to all concerned, provided it comes down by drops, and not by tons together'.

Once the routine of machine-minding was established, the monotony of the work became one of the chief bugbears. This second-phase reaction was well expressed by Ernest Jones, a Chartist leader with a middle-class pedigree, who was influenced by Marx and Engels. The volcano Etna occasionally expired, Jones wrote, but 'man's volcanoes', the steam engines, never rested:

'Women, children, men were toiling,
Locked in dungeons close and black,
Life's fast-failing thread uncoiling
Round the wheel, *the modern rack*.'

And Jones struck the same note as Elliott had done in his famous popular hymn attacking the Corn Laws, 'When wilt thou save the people?', when he went on:

'And a banished population
Festers in the fetid street:
Give us, God, to save our nation,
Less of cotton, more of wheat.'

All technologies generate fear as well as hope, and steam technology not only was no exception, but set a pattern for the future. In disturbing ways of life, it stimulated working-class organisation and led to demands for a social transformation as far-reaching as that which it had itself already created. From early on factory operatives addressed wider issues than the pace of their work or the discipline of the factory. At a mill in Middleton in Lancashire, where calicoes were being woven by steam looms, operatives complained in 1812 that it would be impossible for hand-loom weavers to earn sufficient to support their families if weaving by steam were to become general. Later, some workers objected (like Morris) that the ownership of steam engines concentrated capital in the hands of owner/employers, and became Chartists and socialists, while others turned to the cause of factory reform, seeking as their first objective the statutory restriction of the working hours of factory children in the power-driven textiles industry.

The first of a series of Factory Acts had been passed in 1802, and a further act in 1833 led to the appointment of factory inspectors. Effective factory legislation, however, required continuing agitation and organisation, and later acts were bitterly opposed by employers on economic grounds.

It was during the course of the struggle to secure factory reform and to organise trade unions (the cotton spinners, one of the most important groups in the new labour force, were particularly active in this campaign) that an alternative version of political economy was promoted – very different from that of Ure and people like him. It took as its central tenet the principle of 'a fair day's wages for a fair day's work', systemized by early socialist economists into the doctrine of the worker's right to 'the whole product of labour'.

With support from a variety of sources, including the speeches and writings of the reforming mill-owner Robert Owen, competition was criticised and 'co-operation' proclaimed. Friedrich Engels argued that it was 'the worker who had to bear all the hardships involved in industrial change', and labelled Ure 'the lackey of the bourgeoisie'. Engels found reinforcement for his views in some of the writings of Carlyle, in the evidence of doctors and in official blue books about fatigue and ill health in industrial districts.

Education, of course, was an important factor in the articulation of grievance. Britain had still to develop a public educational system, but schooling needs were met by religious and voluntary bodies, among them the new mechanics' institutes which sought to spread knowledge of science and technology among working men. Some of

At a fete held in celebration of 'winning the coal', Welsh miners and their wives sit down to a pithead tea. On the Rhondda branch of the Taff Vale Railway, which runs alongside the pit, a train filled with the product of their labour is ready for departure.

these institutes had steam power on their curriculum. Conservatives were aware, as the publisher Charles Knight proclaimed, by analogy with steam, that 'Knowledge is power'. There were students in the mechanics' institutes who wanted to learn not only about technology but about socialist political economy as well. Even when they were content to stick to their major task – with the steam engine prominent in their lessons – they could, of course, be accused by conservative critics of threatening to 'uproot society'. For some conservatives steam power itself had controversial connotations. The Society for the Diffusion of Useful Knowledge, founded by Henry Brougham and his friends in 1826, which tried (not without success) to persuade working-class people willingly to adopt new ways of thinking and working, was dismissed in some quarters as the 'Steam Intellect Society'.

Working-class reactions to steam power were not invariably hostile, as one enthusiastic follower of Robert Owen revealed when he described education as 'the steam engine of the new moral world'. There was sufficient adaptation to the new technology to encourage the development of a working-class industrial folklore which took steam for granted. In a broadsheet poem 'The Steam Loom Weaver' the language of steam becomes the language of love:

'The lassie was a steam loom weaver,
The lad an engine driver keen,
All their discourse was about weaving,
And the getting up of steam.'

And so we pass to the inevitable climax:

'Her loom worked well the shuttle flew
His nickers play'd the tune nick-nack,
Her laith did move with rapid motion,
Her temples, healds, long-lambs and jacks . . .
The young man cried your loom works, light
And quickly then off shut the steam.'

There was an earlier version of this poem written in the days of handloom weaving, when the worker had not yet become a 'hand' and when both girl and boy were weavers. The fact that the young man in the later version was a railway engine driver introduces a new character into the literature of steam.

Any account of the gospel would be incomplete if it left out the railways. In Britain the 1840s were a decade not only of social and industrial conflict but of extraordinarily rapid railway development. During this decade, according to contemporary estimates, no less than 600,000 people were involved in railway construction and related activities, roughly as many as were then employed in factories.

Not surprisingly the gospel of steam was revised and extended. First, the application of steam to locomotion seemed to contemporaries to mark the greatest historical divide and the greatest and proudest triumph of steam. As early as 1821, before any steam railway was in regular operation, Thomas Gray produced a pamphlet with the formidable title: *Observations on a General Iron Railway or Land Steam Conveyance; to supersede the Necessity of Horses in all Public Vehicles; showing its vast superiority in every respect, over all the Pitiful Methods of Conveyance by Turnpike Roads, Canals, and Coasting Traders, Containing every Species of Information relative to Railroads and Locomotive Engines.* The pamphlet ran through five editions, and aroused so much discussion that a contemporary remarked: 'Begin where you would, on whatever subject – the weather, the news, the political movements of the day – it would not be many minutes before, with Thomas Gray, you would be enveloped with steam.'

Second, the growth of railways did more to universalize the gospel than any other application of steam. There were a few sceptics such as the French poet Theophile Gautier, who in 1837 branded railways as a 'scientific curiosity, a sort of industrial toy', but most writers were not willing to leave the matter there. 'Man is achieving a victory over time and space,' wrote Charles Knight in 1845, 'but we scarcely know how to contemplate the possible end without something like awe.'

Many poet-journalists turned more with pleasure than with awe to railways, producing some of the worst 'stuffed-owl' verse of the 19th century. One of them wrote of the new type of traveller:

'On iron roads (o'er levelled hills convey'd,
Through blasted rocks, or tunnell'd mountains made)
By steam propell'd, pursues his rapid way
And ends ere noon, what erst employ'd the day.'

A quintessential Victorian poet, Coventry Patmore, described how
'. . . the train, with shock on shock,
Swift rush and birth scream dire,
Grew from the bosom of the rock,
And pass'd in noise and fire,
With brazen throb, with vital stroke,
It went far heard, far seen,
Setting a track of shining smoke
Against the pastoral green.'

The great English novelist Thackeray made the most of the historical break represented by the coming of railways: 'We who lived before railways and survive out of the ancient world, are like Father Noah and his family out of the Ark,' he wrote. 'Your railroad starts a new era, and we of a certain age belong to the new time and the old time.' Novelists could prove just as interested in more general effects of steam. George Eliot, for example, put into the mouth of one of her characters in *Adam Bede* the now well-known lines: 'The world goes on at a smarter pace now than it did when I was a young fellow, . . . It's this steam, you see, which has made the difference; it drives every wheel double pace, and the wheel of fortune along with them.'

Commemorative medals (above) for the first public railways in Britain and France – between Stockton and Darlington and Paris and St. Germain – and a director's ivory free pass, worn on the wrist, provide reminders of the enormous importance of railways when they were the unchallenged pacemakers in land transport.

The message of contemporaries was often optimistic. 'If we can proceed with our railway through the heart of the country,' wrote a Governor-General of India in 1848, 'we shall make rapid strides here both in wealth and stability – for steam here would be the greatest instrument of civilisation for the people.'

Yet in every part of the world the development of railways alarmed people as much as it stimulated them. This was the third implication. The building of railways broke up the unity of cities and disturbed the peace of the countryside, and railway accidents were given more publicity than the ceremonial opening of railway stations. Looking back, the ex-Chartist poet Thomas Cooper remembered a time when

'There was no Rail whereon the steam-steed sped
With snort, and puff and haste to two men pale
With fear, and fill their hearts with distant dread
Of death. . . .'

The death image was powerfully expressed also by the poet de Quincey – 'The rail-cars snort from strand to strand, Like more of Death's white horses' – and by Charles Dickens, for whom the 'conquering engines' of the railway were usually more sinister than friendly: 'The power that forced itself upon its iron way – its own – defiant of all paths and road, piercing through the heart of every obstacle, and dragging living creatures of all classes, ages and degrees behind it, was a type of the triumphant monster, Death.'

The response to technological change was thus very different during the 1840s from what it had been during the 1780s and 1790s. Wordsworth's hope that it would be possible to enlist 'Imagination under the banner of Science' seemed less justified than it had been. The material revolution produced by steam did not bring wealth to all, nor were the implications of the wealth itself universally welcomed. Already there were signs by the mid 19th century of a gulf between the 'two cultures', a divide which ultimately was to seem as wide as that between the two contrasting periods of time – before and after steam.

In the 20th century, surveying the period since the industrial revolution, C. P. Snow in his lecture on the *Two Cultures* chose knowledge of the laws of thermodynamics as the basic kind of scientific knowledge which everyone should share. Literary critics wondered why these laws, drawn up to explain the workings of steam engines, should be regarded as the touchstone of a civilised consciousness. Yet there had once seemed to be poetry in them, not least to the scientists who worked towards their discovery. The French physicist Sadi Carnot wrote in 1824: 'Nature, in providing us with combustibles on all sides, has given us the power to produce, at all times and in all places, heat and the impelling power which is the result of it. To develop this power, to appropriate it to our uses, is the object of heat engines. The study of these engines is of the greatest interest . . . and they seem destined to produce a great revolution in the civilised world.'

The facts of that revolution which took place 'in the civilised world' between then and the end of the 19th century failed to match up to the prophecies. The diffusion of steam power did not succeed in welding 'the individual interests of particular nations into one great universal "public weal" ', and the union of the arts and sciences, so central to the gospel of steam, was in time largely forgotten.

DARK SATANIC MILLS

In Lancashire and Yorkshire, traditional centres of
the textiles industry, mills powered by steam engines
were still being built after the First World War, and
continued to run on steam until long after the Second.
The pictures on these pages, taken during the late
1970s in some of the longest-surviving cotton mills,
record the living history of the world's first
factory system.

*A weaver takes up an end on one of the 500
looms in the Bancroft weaving shed of James
Nutter & Sons , the last cotton mill in the
Yorkshire town of Barnoldswick. The mill,
completed in 1920, was closed in 1978 and
demolished, except for the engine house, in 1980.*

Fronted by a reservoir, the Bancroft mill nestles into a hill, its weaving shed half below ground.

Intensive Labour

Cotton mills have often been called 'dark Satanic mills', though William Blake was not thinking of them when he wrote that phrase in his hymn *Jerusalem*. The mills were certainly noisy, dirty, unhealthy places to work – they had to be kept damp to stop the yarn breaking – and they required a lot of maintenance. Boiler flues had to be cleared, reservoirs dredged, engines stripped. Chimneys were a particular problem. They needed to be regularly inspected, repointed, braced and every ten years waterproofed with boiled linseed oil.

On the roof of the Bancroft shed, workmen whitewash the glass to protect weavers in the humid interior from the greenhouse effect of the summer sun.

A steeplejack ladders a chimney for inspection, hammering in woodchip plugs and iron dogs and lashing on the ladders sent up from below on pulleys.

The 32-foot flywheel of the cross-compound Bancroft mill engine is driven by a small high-pressure cylinder and a larger low-pressure cylinder (right).

Ropes carry the drive from the engine flywheel to all parts of the mill.

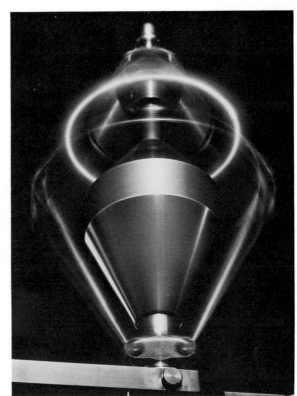

The spinning governor automatically keeps the engine speed to 67 r.p.m.

Glass reservoirs reveal the oil levels in a set of drip-feed lubricators.

Pressure gauges and name plate complete the engine tender's instruments.

A Burnley weaver wearing traditional clogs and apron has her hair neatly done up in a bun to prevent it getting caught in a steam-powered belt drive.

A woman hides her face as she shouts a private message: noise-inured weavers are able lip-readers.   *A finished length of cloth is cut from the loom.*

Cloth carriers take a 54-inch reel of woven cotton for bleaching, dyeing and finishing. These looms could also weave anything from bandage cloth to silk.

A boiler fluer enters the water-bearing part of a still-warm Lancashire boiler to carry out descaling.

In the flue beneath a boiler, men wearing face masks, goggles and lamps rake up the sulphurous ash.

A loom sweeper clears up at the end of the day. By the 1970s night shifts were a thing of the past.

This is the sight that met a weaver arriving for work on a dark winter morning. Thanks to cheap foreign competition and technological change, it is no more.

CHAPTER FOUR
LOCOMOTION

There is nothing so serviceable or so valuable to
mankind as the steam locomotive . . . a machine
easy to make, easy to run, easy to repair, never weary
from its birth in mint condition to the days that saw
it worn, dirty and old; wasteful as nature and as
inefficient as man, very human in characteristic, far
from ideally economic in action but, like our race,
ever in a stage of development, master in emergencies,
its possibilities of improvement inexhaustible.

COLONEL KITSON CLARK, quoted by
L. T. C. Rolt, *George and Robert Stephenson* (1960)

Headed by the 4–4–0 saddle-tank locomotive
Lance, *the first broad-gauge passenger train*
from Penzance pulls into Redruth Station,
Cornwall, on 1st March, 1867. The conversion
of the standard-gauge West Cornwall Railway
(by the addition of a third rail) enabled
passengers to travel all the way from London on
broad-gauge trains, then matchless for speed
and comfort.

The idea of applying steam to travel preceded the work on high-pressure steam which made rapid development of steam loco-motion possible. Indeed, there was as much talk in the middle of the 18th century about improvements in transport as there was about improvements in stationary machines. The *Gentleman's Magazine*, for example, included an article in 1760 on 'a machine to improve inland navigation', three years later an article on 'a machine to row barges against the stream', and six years after that a description (with illustrations) of a propelling machine 'to go without horses'. And in 1768 a gold medallist of the Society for the Encouragment of Arts and Manufactures wrote a paper proposing railroads on which carriages would be drawn by horses or 'by ropes for steam-winding engines'. At the same time, Erasmus Darwin was dreaming of steam-driven 'fiery chariots', and a French army officer, Nicolas Cugnot, actually made one, to travel on the road. A second chariot of Cugnot's with two open-topped high-pressure steam cylinders is preserved in the Conservatoire des Arts et Métiers in Paris.

In 1769 a three-wheel carriage built by the Frenchman Nicolas Cugnot became the first land conveyance to run on steam power. Moving at an unstoppable two to three miles an hour, it ran into a wall and landed its inventor in prison for being a danger on the roads.

Ideas for the future were constantly running ahead of the present. There was even a pamphlet of 1810 by George Medhurst called *A New Method of Conveying Letters and Goods with Great Certainty and Rapidity by Air*. Medhurst's idea was to propel trucks inside a large iron tube by air pressure supplied from stationary steam engines. Later he proposed carriages running in the open which would be driven by a piston inside a tube. (Atmospheric railways of a similar kind, operated by vacuum, were built in the 1840s, but proved impractical.) Another pioneer of atmospheric railways, John Vallance, wrote a pamphlet in 1824 which included an imaginary dictionary entry of the year 1924. Under the heading 'impossible' appeared the definition: 'a word formerly much in use, even among persons of intelligence, but which is now considered to indicate paucity of information, limitation of intellect and the absence of all grandeur of conception'.

All this is in a sense pre-history, yet in the often told story of steam locomotion, it is not easy to say where the pre-history ended and the history began. This is particularly true of railways, the iron horses of the age of steam. Railways, as we have come to know them, have four distinct features, each with its own history – specialized tracks, power traction, organised movements of freight, and regularly timed passenger traffic – and it took time to bring all four features together.

The first tracks may well have been formed accidentally as one vehicle slid into another's ruts; thus they may be older even than the wheel. Wooden-framed tracks, known as tramways, were being laid in collieries in the very early 17th century to simplify the passage of coal wagons; in County Durham one remarkable early 18th-century tramway was several hundred yards long. The idea of using such tramways for other purposes than conveying coal was commonplace by the 1750s.

Power traction had a different history. The first power on the tram-ways was horse-power, and the first steam-driven carriages of Cugnot and his successors did not stick to tracks: they moved on the roads. As for organised movements of freight and timed passenger traffic, these were developed in the golden age of the canal and the stagecoach. Just as the development of water power preceded the development of steam power, so the building of canals preceded the building of railways; and when the first railway carriages were built they were little more than stagecoach bodies on iron frames.

James Watt had specified in his parallel-motion patent of 1784 a number of other improvements of which the seventh was the invention of 'steam engines which are applied to give motion to wheel carriages for removing persons or goods, or other matters, from place to place, and in which case the engines themselves must be portable'. His object in including this clause was protective rather than exploratory: he wanted to forestall other projectors. Because of his distrust of high-pressure steam, he had no immediate intention of pursuing the idea himself. But already, against his wishes, his engine erector William Murdock (remembered today as the inventor of coal-gas lighting) had produced a model steam carriage with a steam cylinder three-quarters of an inch in diameter and driving wheels nine and a half inches in diameter and had run it at six to eight miles an hour. Friends and acquaintances watched it drawing a small model wagon round a room in his house at Redruth in Cornwall. 'I am extremely sorry that W. M. still busies himself with his steam carriage,' Watt wrote in 1786.

Watt's lack of enthusiasm for high-pressure steam was not shared by the Cornishman Richard Trevithick who, when the life of Watt's patent was drawing to a close, began working with non-condensing steam engines, which were nicknamed 'puffers' because the high-pressure steam was exhausted into the atmosphere (rather than into the condenser) after it had done its work.

He had many 'wild fancies', as he called them, about the possibilities of high-pressure steam, though he knew that if it were to be widely employed for 'useful' purposes, much stronger boilers would have to be constructed to withstand the required pressure with safety. Watt, who once said that Trevithick 'deserved hanging for bringing into use the high-pressure engine', was content with boilers as they were: as Dickinson has said, 'he found the boiler merely a hot-water tank and left it very little better'. That there was an element of genuine danger in high-pressure steam was proved in 1803 when a Trevithick stationary engine at Greenwich exploded and killed four people. Debris was thrown two hundred yards, and in the vivid words of a contemporary 'not two bricks [were] left fast to each other either in the stock or round the boiler'.

The change to high-pressure steam seemed in its own time as big as the change from the atom bomb to the hydrogen bomb in the 20th century. There were a number of obvious implications, leaving out the danger. First, vital parts of the existing steam engine could now be discarded – like the condenser and air pump. The engine became far more compact. The great rocking beams disappeared and there was no longer any need for massive engine houses. Second, there were immense possibilities of increased power in 'strong steam' as technology developed, making possible in the 20th century boilers and turbines which generate and use steam at a pressure of hundreds of pounds to the square inch. Third – and this seemed obviously the most important feature – the high-pressure engine could actually move: it was a vehicle, and it saved time while it crossed space. For this reason contemporaries foresaw a fourth implication. 'It is not doubted,' wrote a correspondent from Merthyr Tydfil in 1804, 'but that the number of horses in the

Kingdom will be very considerably reduced.' In fact, because railways generated extra demand for ancillary haulage, this implication was not immediately realized and the number of horses actually increased.

A statue of Trevithick was unveiled in 1932 near the spot on Beacon Hill at Camborne in Cornwall where he travelled by steam carriage on a memorable ride with companions on Christmas Eve 1801. The ride ended in fiasco when the engine overturned and was destroyed by fire. A year later Trevithick took out a patent, which at least one contemporary observer thought would become 'of great national importance', and carried out trials of his 'puffers' at Coalbrookdale (where there were iron rails) and on the roads, including Tottenham Court Road in London. In February 1804 a Trevithick steam engine weighing five tons (with water included) hauled a load of ten tons of iron and seventy men in five wagons for a distance of over nine miles on a cast-iron plate tramway from Pen-y-Daren ironworks to Abercynon Wharf on the Glamorgan Canal in Wales. This was an outstanding achievement, which successfully brought together locomotive and track, although the locomotive was so heavy that it soon broke the cast-iron plates of the tramway.

Although Trevithick did not persist in many further trials, one of his 'dragons' made a spectacular appearance in London in 1808 on a specially built circular railway within an enclosure, to which the public were admitted for an entry fee of a shilling. The engine was called *Catch-me-who-can*, and admission cards for the show carried the motto 'Mechanical Power Subduing Animal Speed'. There is no better example in railway history of the exploitation of the play element in invention, and it was appropriate that the setting should be Euston Square, not far from the spot where the great Doric Arch of Euston Station was to be built when the age of railways had begun. There was a contest element in the early history too. Also in 1808, *The Times* described a steam engine at Newmarket 'now preparing to run against any mare, horse or gelding that may be produced at the next October meeting'. The steam dragon was required to prove itself against the blood-bone-and-muscle horse.

Trevithick continued to indulge in 'wild fancies' – some involving the use of steam in armament for the war against Napoleon – but he went bankrupt in 1811, and after spending ten often difficult, often exacting, years trying to make money in South America, he returned to Britain to find that other men had developed high-pressure steam as a working proposition.

By then, the scene of the story had shifted from the South-West and London to the industrial North-East of England. Northerners were more concerned with economic necessities than with wagers on horses and engines, and they included one man of genius and perseverance, George Stephenson, who was soon to be described by Lardner, Smiles and others as the 'Father of the Locomotive'. Stephenson, who started with no advantages of birth and worked first as a brakesman at a pit, had been given charge in 1812 of all the machinery in the collieries of the so-called Grand Allies, England's biggest mining consortium. He was not the first man to produce locomotives in the North of England, however, since in that very year a Leeds man, John Blenkinsop, working

A ticket for Richard Trevithick's circular-railway show in London in 1808 has been inscribed by one of the Cornishman's enthusiastic followers. 'A bet was made,' he wrote, 'that this machine would go farther in 24 hours than a race horse'. In fact a rail soon broke and the engine overturned.

George Stephenson,

A stevengraph, woven in silk in about 1880, commemorates the achievements of George Stephenson. He appears with his Locomotion, Rocket *and another locomotive of the Liverpool and Manchester Railway, the first public inter-city line for both goods and passenger traffic to be exclusively operated by steam.*

William Hedley's Puffing Billy *steams on the Wylam Colliery Railway, Northumberland, in about 1860. Built in 1813, it was withdrawn and preserved at the Science Museum, London, in 1867. It is the oldest locomotive in existence.*

with an ironfounder, Matthew Murray, introduced a regular steam service (with two-cylinder locomotives called the *Prince Regent* and the *Salamanca*) at the Middleton colliery tramway near Leeds. (There had been a wooden track at Middleton as early as 1758, and soon four Blenkinsop locomotives displaced fifty haulage horses.) There were other challengers too, notably William Hedley, a colliery superintendent of Newcastle; his engines *Puffing Billy*, now in the Science Museum, and *Wylam Dilly* took to the tracks in 1813.

Stephenson was the first man, however, to make and run a loco-motive with flanged wheels on a track laid with cast-iron rails. His first engine, *Blücher*, an advance on any earlier engines, ran in 1814, the year before the real Prussian Blücher rushed to the Duke of Welling-ton's assistance at the Battle of Waterloo. If Trevithick's success at Pen-y-Daren was the first great landmark in the history of the railway, this was the second. Stephenson's engines were remarkably well made, and seven years later one of the first great railway enthusiasts – there was soon a host of them – was to describe them as 'superior beyond all comparison to all the other engines I have ever seen'.

The third landmark was the opening of the Stockton and Darlington Railway, the first public railway to employ locomotives, on 27th September 1825. Stephenson's engine *Locomotion* hauled the inaugural train and is now preserved at a railway museum housed in the original Darlington Station. It is important, nonetheless, not to exaggerate the significance of this event as far as the history of steam power is concerned. Horse traffic on the new railway was given more publicity than steam-driven engines by the company, which even showed horse-drawn carriages on its seal. Until 1833 steam locomotives were used to haul only the freight trains.

The fourth landmark was the opening of the Liverpool and Manchester Railway on 15th September 1830. Following the Rainhill locomotive trials, organised by the directors of the railway the previous year and convincingly won by Stephenson's *Rocket*, it had been decided that all trials, organised by the directors of the railway the previous year and a much described occasion, was marred by the death of one of Britain's leading politicians, William Huskisson, who was hit by a train. Nor was this the only incident. An angry mob gathered at Manchester Station, where demonstrators were more concerned with airing political griev-ances than with welcoming progress. Nonetheless, it was proven that there was progress – and progress without precedent – when Stephenson himself carried the dying Husskisson fifteen miles in his *Rocket* in twenty-five minutes – an average of nearly forty miles an hour. 'This incredible speed burst upon the world,' Samuel Smiles wrote later, 'with the effect of a new and unlooked for phenomenon.'

Most early travellers by railway were impressed above all else by the speed of travel, 'really like flying' in the words of the actress Fanny Kemble, who treated the locomotive as 'a snorting little animal' which she was 'rather inclined to pat'. There is an unforgettable passage in a letter of Carlyle's, describing his journey from London to Preston in Lancashire in 1839: 'To whirl through the confused darkness, on those steam wings, was one of the strongest experiences I have ex-perienced – hissing and dashing on, one knew not whither . . . we went over the tops of houses . . . *under* the stars; not under the clouds but among them. Out of one vehicle into another, snorting, roaring, we flew: the likest thing to a Faust's flight on the Devil's mantle.'

Very soon railways in Britain became routine, with the freight traffic being more significant economically, if not socially and culturally, than the all too articulate passengers. Between 1825 and 1837, the year when Queen Victoria came to the throne, 500 miles of railway track were opened. By 1844 there were 2,000 miles of track and by 1852 over 7,500 miles. It became possible to talk without exaggeration of a railway system with regular *Bradshaw* timetables, then as rigid as the timetables of the mill or the warehouse.

Locomotive building became an industry, and Robert Stephenson, George's son and close partner in many of his enterprises, soon proved himself outstanding as a pioneer, making improvements in each successive engine design. Another engineering dynasty, the Brunels, developed the broad-gauge Great West Railway, on which Daniel Gooch's *Firefly* set a new speed record of fifty miles an hour on a journey from Twyford to Paddington in 1840. (Another Gooch engine, the *Great Britain*, set up a record of sixty-eight miles an hour from Paddington to Didcot eight years later.) Swindon became the first locomotive depot of the Great Western, one of a number of railway towns, among which Crewe in Cheshire, the locomotive depot of the Grand Junction Railway (later the London and North Western), was the most important.

Breaking records did much to popularize steam, but it did not guarantee business success. The important war of the gauges was won not by the Great Western but by the standard gauge party, backed by all the other railway companies. The victory came despite the comfort of broad gauge carriages and the speed and power of the magnificent Swindon locomotives with their huge boilers and exceptionally high pressure of 100 pounds to the square inch.

During the early years of railway building in Britain, investment often turned into speculation, and the speculation reached for a time the dimensions of a mania. Because the whole process of railway building in Britain was rapid and difficult, the number of financial interests involved so great, and the technology so new (involving as it did, dramatic feats of civil engineering as well as advances in mechanical engineering), British railways cost more to build per mile than any later railways in any continent. They cost twice as much, for instance, as German railways were to cost, and four times as much as railways in the United States. British railway building made some men famous and others rich (and some both), among them Joseph Locke of the Grand Junction, and 'the Railway King' George Hudson, who is commemorated in York, the old cathedral city which became another of the great railway centres. Some companies, however, had a very troubled history from the start, with the Oxford, Worcester and Wolverhampton, for example, meriting the nickname it soon received: 'Old Worse and Worst'.

Because of the lead Britain had won in railway development, British contractors and labourers went on to build railways everywhere – in Europe, Asia, Africa and America. One of the greatest contractors was Thomas Brassey who, after constructing 1,700 miles of railway in Britain, became an expert in railway construction overseas. His entry in the *Concise Dictionary of National Biography* speaks for itself: Canadian Grand Trunk (1852–59), Crimean (1854), Australian (1859–63),

Argentine (1864) and Indian (1858–65). This imposing list leaves out France, where he, Joseph Locke and their men built the railway from Paris to Rouen and then on to Le Havre during the 1830s.

As early as 1823, railway track had been constructed in mining districts of France, with horses employed to provide the power, and steam locomotives were used in 1832 on the line from St. Etienne to Lyons. Nonetheless, there was opposition to the railways, particularly if they were privately constructed: the poet Lamartine compared railway barons with feudal barons, and another poet, Alfred de Vigny, urged that people should use them only in case of illness, death and war. The railway network was slow to develop. In 1841 France had only 350 miles of track. Ten years later, however, the establishment of the Second Empire of Napoleon III changed official attitudes, and the 1850s became the peak years of construction in France. By the end of the century there were over 28,000 miles.

In locomotive building the French played a more distinguished role, pioneering compound locomotives of two, three and four cylinders in the 1870s and 1880s. Double expansion in a compound engine involves expanding steam first in a high-pressure cylinder and then passing it to a low-pressure cylinder of larger dimensions where it expands further. Compound steam locomotives had higher construction costs, but they were more economical and they tended to last better. Four-cylinder compounds were soon taken up on all the French main lines, and were also adopted in many other European countries. In the 1930s their performance was brought to a peak of perfection by André Chapelon, who has been described as the greatest locomotive engineer after Robert Stephenson.

Neighbouring Belgium provided a sharp contrast with France as far as railway building was concerned. Belgium had a railway system even before Britain – and an articulate and consistent government policy towards railways, which were mainly state-owned and controlled. The first railway, the line from Brussels to Malines, was completed in May

On the Liverpool and Manchester Railway, first-class passengers were provided with closed carriages while the second-class travelled in the open. Already by 1831, when this print was published, a regular mail coach was attached to the rear of the first-class trains.

At Leicester Station in the days of gaslight, the ticket collector was responsible for changing the train indicators and setting the departure clocks. In the 1830s the growth of regular train services had ushered in the age of timetables, and in particular of Bradshaw's Guide, a major text in the gospel of steam.

1835 and carried over half a million passengers in its first year, more than the whole of British passenger traffic at that time. Dionysius Lardner in his invaluable study *Railway Economy: A Treatise on the New Art of Transport* (1850), which complemented his earlier general treatise on the steam engine, extolled the element of deliberate planning in Belgium under the direction of 'a special railway committee, invested with adequate powers'. In addition, Belgium offered the world in 1843 one of the most grandiloquent railway poems, 'Harmonie' by Theodore Weustenraad. '*Symbole intelligent de force créatrice*' ('Intelligent symbol of creative power'), Weustenraad began his invocation to the railway, continuing with references not only to its 'vigilant arm' but to the way in which it served 'the great work of God'.

In divided Germany a number of local lines developed at a very early date, the first of them a track built in 1835 between Nuremberg and Fürth and operated by Stephenson engines. A number of German writers saw very quickly how important railways could be in any programme of German unification: the economist Friedrich List, for example, wrote as early as 1833 of the need for 'a general German system'. By the start of the 1850s, when Napoleon III developed his railway policy in France, about 3,000 miles of narrow-gauge line were open in the area of what after 1870 was to be the German Empire, and by 1900 total railway track had increased tenfold.

German railways were constructed more cheaply not only than British but also Belgian and Dutch railways. Land was cheap, as in the United States; and in Lardner's words, 'The vast expenditure for earth work and costly works of art, such as viaducts, bridges and tunnels, by which valleys are bestridden and mountains pierced in the English system . . . was not attempted.'

There were of course parts of Europe such as Switzerland where mountains had to be pierced, and some of the most interesting European railways were mountain railways, constructed during the last twenty

years of the 19th century. In Switzerland special brakes had to be provided for engines and carriages, and there were Alpine travellers who found it almost as exciting to travel by train as to climb mountains. It was at the end of the 19th century too, that international links were completed between Calais and Constantinople. These trans-continental railways soon had their connoisseurs. The Orient Express from Paris to Constantinople had not yet acquired the aura of romance which it was given through novel, film and music between the two great 20th-century wars, yet there was a touch of romance in the complex diplomacy leading up to the inter-state agreement of 1883 which made its opening possible. The Austrians completed their section first, in 1884, but the Bulgarians, eventually using English rails, and the Turks, proved very

A steam railcar, working on the Locher rack and pinion system, grinds its way up the three-mile track of the Pilatus Railway in the Swiss Alps. Opened in 1889 and steam-powered until 1936, the railway is the steepest in the world to be operated by locomotives, with a gradient of 1 in 2 for most of its length.

slow. One contentious dream was for a railway from Berlin to Baghdad, which would expand a German sphere of influence in the Middle East. Samuel Smiles directed attention to many of the early international developments and the ideas behind them in the introduction to a revised edition of Volume V of his *Lives of the Engineers*, which appeared in 1879, dwelling particularly on the United States, which in a vivid phrase of the historian Daniel Boorstin, was very soon about to become a land of 'mobile people and magic machines'. In its continental expanses the railway had an independent and absorbing history, although the first American application of steam to transport was not the locomotive but the steamboat, described later in this chapter. During the 1830s the powerful state of Pennsylvania built canals rather than railways to link up with the developing West.

American engineers had bold ideas for developments in transport. 'The time will come,' wrote the great engineer Oliver Evans in 1812, 'when people will travel in stages moved by steam engines, from one city to another, almost as fast as birds fly. . . . A carriage will set out from Washington in the morning, the passengers will breakfast at Baltimore, dine at Philadelphia and sup at New York, the same day.' Another engineer, John Stevens, went further. 'I can see nothing to hinder a steam engine from moving with a velocity of one hundred

'A Close Finish' is how Henry Alken described this trial between mechanical and animal speed in 1866. Steam and horses never mixed well at any speeds.

miles an hour,' he wrote, also in 1812. Evans and Stevens were inventors as well as prophets. But neither man clearly foresaw the chronology of development. Evans continued to dream (as did a number of British pioneers) of using steam carriages on 'the common road', and Stevens did not understand the complexities either of civil engineering or of railway management.

The first railway proposal in the United States came in 1813 – the building of a line between Philadelphia and New York – but it was not until after the Liverpool and Manchester Railway's Rainhill trials in 1829, which were attended by Horatio Allen, the chief engineer of the Delaware and Hudson Canal Company, and E. L. Miller, resident of Charleston, South Carolina, that the railway became a practical proposition in the United States. Both Allen and Miller ordered locomotives during or after their visit, and the first two British locomotives to be imported into the United States were the *America*, built by the firm of Robert Stephenson (and costing $3,663.30), and the famous *Stourbridge Lion* (cheaper at $2,514.90), which operated on a railroad built exclusively for commercial traffic at Honesdale, Pennsylvania.

Among the earliest railways there were straightforward place-to-place business projects conceived on similar lines to those undertaken in Britain. The greatest excitement of American railway development as

Railway Mayhem

The dangers of railway travel and the disruption of county pleasures provided themes for cartoonists from 1830 on. Boiler explosions in particular (bottom right) offered rich opportunities for the drawing of severed limbs and the exercise of heavy irony.

The 'Probable Effect of the Projected Rail-Road to Brighton' was thought to be a lowering of tone.

This was Punch's *view of 'How to Insure Against Railway Accidents'.*

Bits of body fly about in a cartoon on 'the Inconvenience of a Blow-up'.

compared with British, however, was that the railway not only connected established places like Philadelphia and New York, but brought new places into being. Among these was Chicago, which became one of the great new cities of the world. The railway pushed out, too, into the unknown over huge continental distances, completely changing patterns of expectation as much as patterns of daily life. 'That land wasn't worth two cents,' a Congressman said, 'until the railroad came.'

As in Europe, the railways lowered the cost of moving freight, made it possible to move perishables over greater distances, thereby extending markets, provided a simpler and cheaper system of passenger traffic, and speeded up and unified postal communications for private, public and business purposes. In the United States, as in Germany and other countries, railways also had a political effect. Stevens had foreseen that steam would unite 'these states' into 'one family internally connected and held together in indissoluble bonds of union', and not even the extensive and often imaginative military use of railways during the American Civil War took the power out of this long-term prophecy.

During the 1850s railway mileage in the United States increased three times, with the help of land grants by the States (the Illinois Central Railway received the first such grant in 1851). The first transcontinental line was completed in 1869, when the United Pacific Railroad Company, building west from Omaha, Nebraska, met the Central Pacific, building east from Sacramento, at Promontory Summit, Utah. The second, the Southern Pacific, followed in 1881, and there were two more in the next year: one in the north (which received no land grant subsidies) and one in the south, the Santa Fe, which had begun with the relatively modest original objective of monopolising the trade of mining companies (another link between mining and steam). Ten years later the Great Northern was completed. During the 1880s as many miles of track were laid as in all the years from the beginnings of railway development in the United States to 1870. This was spectacular progress, and prompted a lyrical American novelist to claim that 'there is more poetry in the rush of a single railroad train across the continent than in all the story of burning Troy!' By 1920 the mileage of American railroads exceeded that of the whole of Europe and constituted one third of the world's total.

There was much else, however, besides poetry in the story: intrigue, intimidation, corruption and conflict. There were also sharp technical contrasts with the history of railways in England. American railway tracks were different in appearance and lighter than those of Britain. They were often single and 'snake-like'; expenditure on the smoothing of curves, the levelling of hills and the cutting of tunnels was usually kept to a minimum. 'Any saving in the cost per mile of a railway,' wrote a transport expert, Captain Douglas Galton, in 1857, 'adds to the means available for extension; and in a rapidly developing new country capital is dear.'

Of course, there was a price to be paid. Accidents were frequent, and they were widely publicized. 'Biggest and fastest' was the motto, not 'best and safest'. After one terrible accident in 1858, *Harper's Weekly* wrote: 'Nobody's murder. The railroads are insatiable. The boilers are bursting all over the country. . . . Human life is sadly and futilely squandered – but nobody is to blame. Boilers burst themselves, rails

Men of the Central Pacific and Union Pacific Railroads celebrate the completion of America's first transcontinental rail link at Promontory Summit, Utah, in 1869. The directors shook hands and hammered in a golden spike (left), inscribed with their names and the words: 'May God continue the Unity of Our Country as this Railroad unites the two great oceans of the world.'

break themselves. And it may be questioned whether the consequent slaughter of men, women and children is not really suicide.'

State governments – and Congress – gave away land and privileges, often lavishly, before they enforced even the simplest regulation, and some of the individual railway developers acquired great private power. There was an important change in 1887, when the Inter-State Commerce Act placed railways under Federal regulation, yet despite the activities of the newly founded Inter-State Commerce Commission, the railway companies – and a few individuals who dominated some of them – continued to disregard both law and opinion. It was only during the first decade of the 20th century, when the railway network was still expanding, that effective federal regulation was established.

American locomotives differed from British ones partly because of the differences of track. Light and often sharply curved American-built track did not easily bear English-built locomotives such as Stephenson's *Stourbridge Lion*, and although Stephenson tried again to meet American operational requirements with his *John Bull* (1831) and his *Davy Crockett* (1833), as did other British locomotive builders, American builders soon developed their own locomotive styles in the interests of both low cost and dependability. The indigenous locomotive building industry grew with remarkable speed. Ten firms were established in 1840, and the number grew to as many as forty-two before rationalization and specialization cut the figure to six by 1900. In 1840 fewer than 100 locomotives a year were being produced, but this figure had multiplied ten times by 1880 when the total number of locomotives in use had passed the 15,000 mark. As early as 1838 American locomotives, built by Norris of Philadelphia, were being exported to Europe.

The main lines of American technical development were dictated by the attempt to secure greater power through the increase of boiler and cylinder size and the raising of steam pressure. Already by the late 1830s pressures of 120 to 130 pounds per square inch were being raised in the United States – double those of the early English locomotives.

American locomotive design did not remain constant during the 19th century. There was a vogue in the 1830s and 1840s for 'grasshoppers', small locomotives with vertical boilers, and during the 1860s for powerful 'moguls' with 2–6–0 wheel arrangements, the first of which were made for the New Jersey Railroad and Transportation Company. Compound engines, including a four-cylinder compound built at the great Baldwin Locomotive Works, began to be used increasingly during the 1890s. There was also a trend towards standardization and use of interchangeable parts. Certain features of American locomotives soon became distinctive: 'trucks' or 'bogies' under the forward end of a sharply tapered boiler; 'equalizing' suspension bars by which the weight was distributed between all the wheels, whatever the position of the engine or the state of track; 'cowcatchers' in front of the train to remove obstacles from the track; 'spark arresters' and sand-boxes; and powerful headlamps, bells and heavy whistles.

Poets, novelists and artists loved or hated the whistle:

'Down in the meadow
Meadow so low,
Late in the evening,
Hear the train blow,'

went a popular refrain. In his picture *The Lackawanna Valley* the painter George Inness included a puffing train which caused no disturbance: railway and landscape were one. Yet for the novelist Nathaniel Hawthorne, who questioned the association of railways with progress in his sketch *The Celestial Railroad*, the whistle broke all the peace of America's 'sleepy hollows'. 'Hark! There is the whistle of the locomotive – the long shriek, harsh, above all other harshness, for the space of a mile cannot mollify it into harmony. It tells a story of busy men, citizens, from the hot street, who have come to spend a day in a country village, men of business; in short of all unquietness; and no wonder that it gives such a startling shriek, since it brings the noisy world into the midst of our slumbrous peace.'

Walt Whitman, who did not much like 'sleepy hollows' or 'slumbrous peace', formed a quite different judgment. 'The strong quick locomotive' had helped to create the America he knew, continental America, teeming with life. In his poem 'To a Locomotive in Winter' he wrote of its 'black, cylindric body, golden brass and silvery steel', the 'tremendous tinkle' of its wheels and its 'great protruding headlight fixed in front'. There was 'fierce-throated beauty' here and 'lawless music'. The engine was the 'type of the modern – emblem of motion and power – pulse of the continent'.

There was far more railway folklore in the United States than in Britain. 'Lord, Lord, I hate to hear that Lonesome whistle blow' provides a prelude, as do the less well-known lines 'Every time a freight train makes up in the yard, Some po' woman got an achin' heart'. 'Working on the Railroad' became a famous song in one generation, 'Chattanooga Choo-Choo! Won't you Carry me Home?' in quite another. The first commercial film to be made was *The Great Train*

Above the busy horse-drawn traffic of 1870s New York, trains pass on the Elevated Railway at Franklin Square, guarded from derailments by wooden check rails. In London, which adopted the opposite solution to urban congestion (right), passengers take a trial run on the first Underground between Paddington and Farringdon in 1863. Surface condensers helped take care of the steam, but the smoke had to take care of itself.

Robbery. Steam played a major part in family history as much as in national history as Americans moved their locations and their destinies. It provided hope for some communities and despair for others, left 'off the track'. It supported tycoons in extravagant lifestyles and labour gangs (including Chinese coolies) in the roughest of conditions on the opening frontiers. It drew its shareholders, shrewd or gullible, from all sections of the urban community, and forced farmers into resistance against unfair practices, prompting the formation of the Granger movement, the first important farmers' organisation in American history, during the 1870s.

The assessment of the economic effects of American railways has generated as much excitement in academic circles in recent years as the shock of the political effects once did in radical circles. Indeed, it was the effort by a few American economic historians to quantify the consequences of railway building and use which prompted the publication of some of the earliest examples of quantitative economic history, soon called by them 'Cliometrics'. R. W. Fogel, who was to advance the whole subject, sought, as did other historians, to get behind the rhetoric of steam and to measure as exactly as possible 'the social saving' of railroads. After comparing railway costs and benefits with those of alternative methods of transport by road and water, he concluded that the two main benefits achieved by the railroad were the reduction in inventory or storage costs, and the reduction in less efficient wagon transportation.

Many of the same railway themes that we can trace in the United

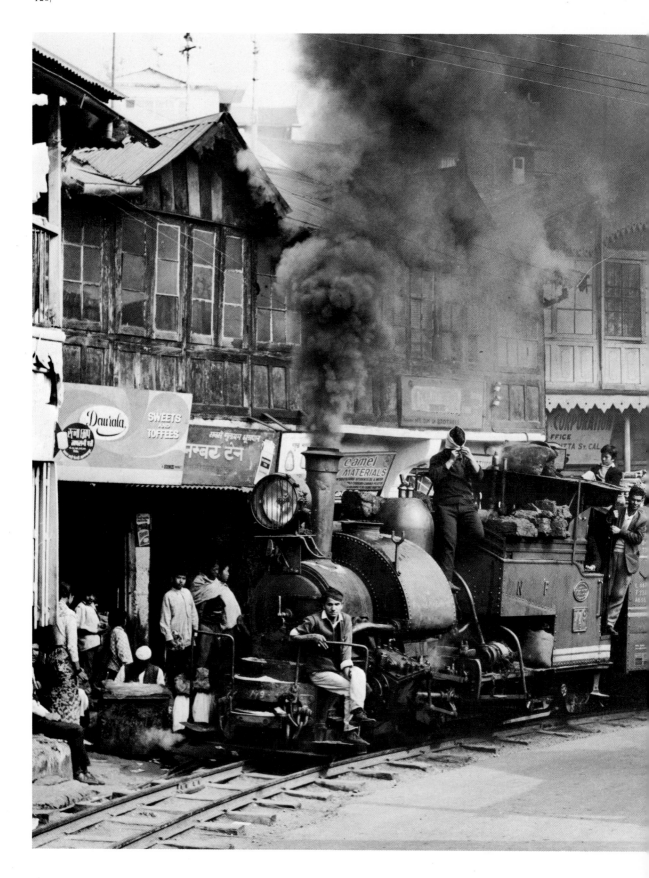

Encumbered with passengers and fuel (left), a Class B tank locomotive built in 1892 by Sharp, Stewart & Co. of Glasgow pulls a train up the main street of Kurseng, a town on the Darjeeling Himalayan Railway. The engine was still keeping alive the British tradition of steam power in India more than 70 years after being supplied.

In Africa and China the opening of new railways had an imperial flavour. Cecil Rhodes's dream of a railway from the Cape to Cairo was clearly emblazoned on Engine No. 1 (below) before it took the first train from Umtale to Salisbury, and the arrival of the first locomotive in China, suitably assisted by coolies (bottom), was seen as a milestone in the spread of 'remunerative railways in developing countries'.

States can be traced in all large developing countries, beginning in the American north with Canada. 'I am neither a prophet, nor the son of a prophet,' a Canadian speaker told a meeting in 1851, 'but I believe that many in this room will live to hear the whistle of the steam engine in the passes of the Rocky Mountains and to make the journey from Halifax to the Pacific in five or six days.' At that time, although the first Canadian line had opened in 1835, and the first train had run in 1836, there were still fewer than a hundred miles of Canadian rail operating, and development remained slow. Ten years later the railway mileage had reached only 2,000. British speculators in the Grand Trunk, incorporated in 1852, were expecting quicker returns than those which they were receiving, and the trans-continental route to the Pacific was still a dream which had its moments of nightmare. Indeed, one Canadian government fell in 1873 after a 'Canadian Pacific' scandal.

British Columbia stipulated the building of the Canadian Pacific route as a condition of entering the Canadian Federation in 1871, and the line was completed in 1885. Two other lines reached the Pacific coast in 1914 and 1915. 'Only the steam locomotive could have conquered Canada,' David Morgan has written in his introduction to *Canadian Steam* (1961), and it is not surprising that great Canadian historians, like Harold Innis, have put communications – from railway to radio – at the centre of their studies, preparing the way for Marshal McLuhan and his aphorism 'the medium is the message'. If American interest in 'Cliometrics' owed much to the study of the railway (as did detailed British studies of 19th-century governmental processes) so, too, did Canadian cultural sociology, and all those have influenced research in other countries.

The difficulty of keeping open the Central Pacific Railroad over the Rockies in winter is well illustrated as a locomotive backs down the track pulling a giant block of compacted snow. The railroad company employed thousands of men to dig trenches on either side of the line and cut out the obstructing snow so that it could be disposed of chunk by chunk.

In Latin America, where railways posed equally difficult technical problems, they played a part in economic development, particularly in Mexico where 15,000 miles of railway were built to market new mineral wealth; in Peru, where the 'nitrate lines' represented the most extensive system of railways in South America; in northern and southern Chile; and in Brazil, where trains hastened the early development of both the coffee trade and the city of São Paolo. Yet in this respect São Paolo was exceptional, since railways did not generally stimulate urbanization (or political unification) in Latin America, and there were large parts of the continent untouched by rail, where mules were not supplanted by locomotives. At least one railway, the Madeira-Mamoré line, was said to go 'from nowhere to nowhere'.

The chief figures in the story of Latin American railway building were foreigners, some of them remarkable characters, like the Englishman Henry Meiggs, 'the railway king of Peru'. It was they who took the key decisions. Although there were groups of 'steam enthusiasts' in Latin America, there was a sense in which the railways did not ever really belong to Latin Americans.

The most dramatic episode in Latin American railway history was the crossing of the Andes. Feasibility studies for a route between Buenos Aires in the Argentine and Valparaiso in Chile were made in 1886, but only 57 miles of the line were complete by 1891, and there was a four-year building lull between 1895 and 1899. The mountains offered both severe physical and political complications, and the Mendoza River, which was always changing its channel, had to be given an artificial embankment before the railway could be completed. The towns of Mendoza on the Argentinian side and Los Andes on the Chilean side were only 156 miles apart, but the railway between the two had to ascend 11,500 feet. There were problems, too, with the locomotives. German-built 'rack adhesion locomotives' were the first to be used, with British Kitson-Meyers taking over later. These were 'special types' and special steam locomotives invariably meant trouble. Only electrification was to solve the operational problems.

There was the same patchy story also in the continent of Africa, where foreigners were even less concerned than they were in South America with local susceptibilities. Railways were part of an imperial game in which the quest for profit and power influenced the developing network. 'The railway is my arm and the telegraph my voice,' the imperial pioneer Cecil Rhodes once exclaimed, while the Boer leader President Kruger, hostile to foreign economic interests, stated equally firmly that 'every railway that approaches me I look upon as an enemy on whatever side it comes'. There were no fully independent black states except Ethiopia, where the Emperor Menelik authorised the building of a railway from Addis Abbaba to Djibouti in 1892. It was not completed until 1917, long after his death, and the early stages of its history were dominated, as was the history of other railway projects in Africa, by fierce rivalries between British and French interests.

On the Union Pacific Railroad, a locomotive shoots steam to clear the track of buffalo, which frequently impeded traffic in the early days of rail travel across the prairies.

Egypt and South Africa at the extremes of the continent were the two countries which benefited from railways most. Egypt secured its first railway in 1834, before Sweden or Japan, and by 1914 had more miles of railway than the Ottoman Empire. In South Africa the first Cape Colony line was built at the end of the 1850s, and the discovery of Kimberley's diamonds and Witwatersrand's gold provided an added incentive to build new lines.

Elsewhere in Africa railways played an important part in opening up the continent in war and peace. The Ashanti War in the Gold Coast prompted the building of a line from Takoradi to Kumasi during the 1870s. In the first decade of the 20th century the extension of the line enabled exports of gold to be multiplied seventy times and promoted the rapid expansion of cocoa production for overseas markets. In the Belgian Congo, where the first railway was started in 1889, the annual expenditure on the line from Matadi to Leopoldville, vital to the economic exploitation of the country, for several years exceeded that of the colony itself.

Strategic as well as economic interests played a major part in the extension of railways in Russia, China and India. The first Russian railway, the St. Petersburg–Pavlovsk, was completed in 1836, and used horse traction until its official opening a year later when British locomotives were introduced; but not until 1848 was a line of any length built by the Russians: it carried troops from Warsaw to the Austrian frontier to crush the Hungarian rising of that year. The important inter-city line between St. Petersburg and Moscow was built three years later with a five-foot gauge which became standard throughout Russia. It was solidly constructed, unlike other Russian lines such as the Moscow to Kursk Railway, which acquired the nickname of 'the Bone-Breaker' on account of frequent accidents. There was no railway boom until the 1860s. By this time the reforming Tsarist Government had started selling off state railways to private interests and the number of private companies was increasing, but this policy was reversed as the importance of railways was more generally recognised. Total mileage increased more than seventeen times between 1857 and 1876, stimulating the grain trade and opening up new parts of the country. By 1890, more than 20,000 miles of track existed and ten years later 30,000 – and there was record annual building in each of the four years from 1898 to 1901. Freight traffic, which had doubled between 1880 and 1890, doubled again in the last decade of the century.

Many of the locomotives and freight trucks were produced in Russia, at first at the famous Alexandrovsky Works near St. Petersburg, which were taken over by the State in 1851. Some locomotives had a rare 6–4–0 wheel arrangement, and the impressive-looking engines of the 1860s betrayed a strong American design influence with their large spark-arresting chimneys. The first engines had no cabs for their drivers, but during the 1870s at least one Tsar, Alexander II, took enough interest in the welfare of locomotive drivers to issue a decree that the platforms of all locomotives should be railed in to prevent men falling off when the surfaces were iced over.

Two particularly exciting ventures in Russia were the Trans-Siberian and the Trans-Caspian Railways across the Steppes to Samarkand. The Trans-Caspian, completed in 1888, was a largely military undertaking,

In the southern borderlands of Russia, an officer and his A.D.C. alight from a Trans-Caspian Railway guard train with double-decker carriages. According to the French account in which this engraving appeared, the train had a restaurant car and kitchen car for the officers; three kitchen cars for the other ranks (three companies of 200 men each); and a hospital car. When the inaugural train ran on the Turkestan-Siberian Railway, completed in the 1930s (right), the military were no less in evidence.

designed to facilitate the movement of troops to Russia's southern borders rather than to convey civilians from one city to another. It is worth noting that the railway station at Bokhara was ten miles outside the town, and that at Samarkand five miles outside. 'This odd practice,' a British writer observed, 'is entirely a Russian habit, for strategy is more desirable than the convenience of travellers.' As much as any railway in the world, the Trans-Caspian fulfilled Ebenezer Elliott's prophecy in 'Steam in the Desert', crossing 200 miles of sand between the Caspian Sea and the Oxus River.

Politics and strategy also played an important part in the building of the Trans-Siberian. The first section was opened in 1896, and six years later the line was complete from Moscow to Vladivostock. It had been built one-and-a-half times faster than the Canadian Pacific, largely because work started simultaneously on different sections, labour costs were low and the capital costs of bridges, loops and watering points were carefully controlled. Yet the line required many outstanding engineering feats such as the building of twenty-eight tunnels in a stretch of forty-two miles.

The chief advocate of the railway was Count Sergei Witte, one of the most important Russian Ministers of the *ancien régime*. A railway enthusiast, he used every kind of argument – strategic, political and economic – to support the venture. He had started his career in railway administration and in 1889 had established a railway department in the Ministry of Finance. Witte believed that railways would strengthen Russian imperialist interests, but Marxist historians saw railway development in a different light. As M. N. Pokrovsky was to write after the Russian Revolution, 'Landed property calls the railways into existence, in order to reach the wide and most profitable European market. The railway gives birth to metallurgy, metallurgy creates the most revolutionary section of the proletariat, which buries the ancestor of the whole system – landed property.'

In India and China the coming of railways brought controversy. British civil servants were quick to point out the advantages of railways in India: one Commissioner reported in 1846 that they were a 'desideratum' and could be as profitable in India as in Europe; and the far-sighted Governor-General, Lord Dalhousie, wrote a forceful Minute of 1853 laying down future policy, one year after the first locomotive had travelled from Bombay to Thana. Another Britisher, Sir Malcolm Stephenson, wrote an impressive study of Chinese railways in 1864 in which he pleaded for the introduction of a 'comprehensive system decided at the outset' which would 'avert the evils of the English want of such a system, where in many cases double capital has been laid out to perform work which one expenditure would have adequately provided for'.

British views were not accepted easily by all influential Indians or Chinese. Indian critics of railways – and at least one British Governor-General – wanted irrigation schemes to be given priority, and some of the first travellers were very sceptical about the operation of the railways. A newspaper of 23rd August 1854 described the suspicious reaction of an Indian trader who, after arriving by train at Hooghly in Calcutta, doubted whether he had been brought to the right destination and went down the street asking people the name of the place. By the end of the century, however, when 25,000 miles of track were open, Indians had come to take great pride in both their locomotives and their railway stations. The imposing Victoria Terminus in Bombay, built in 1888 and officially opened on Queen Victoria's Jubilee Day, was not simply an imperial landmark. (Nor was the statue of a female figure at the front of the building Queen Victoria; it represented 'Progress'.) Like railway stations and junctions throughout India, Victoria Terminus became a popular Indian *rendezvous*, with its own busy local population serving the needs of the travellers.

In China acceptance of 'progress' was far slower. Chinese bureaucrats were always jealous of foreign influence, and local Chinese feeling sometimes caused delays. The first railway project – a line from Shanghai to Woosung – had to be temporarily dropped in 1865 after there had been objections to destroying graves in the line of the track. When the line was eventually opened in 1876, the first Chinese passengers were allowed to travel free. Such concessions did not wholly allay Chinese hostility, and later in the century there were threats of anti-railway riots. By 1900 there were still only 250 miles of railway, but concessions had been signed with Western powers authorising the building of another 5,000 miles, including the Peking to Nanking, the Canton to Kowloon (linking China with Hong Kong), the Peking to Hankow (to be constructed with French and Belgian capital), the Chinese Eastern Railway across Northern Manchuria, and the Southern Manchurian Railway to Port Arthur.

In another country of great continental distances, Australia, the first railways were two very short lines in Victoria and Sydney, built in 1854 amd 1855. The first long-distance line from Melbourne to Sydney was not opened until 1883, handicapped even then by differences of gauge which prevented the development of a true inter-city system. New South Wales had a 4 foot $8\frac{1}{2}$ inch gauge, Victoria and parts of South Australia a 5 foot 3 inch gauge, and the remaining Australian colonies, including Queensland, Tasmania and Western Australia 3 foot 6 inch. There were gauge differences, too, in New Zealand, where the first railway – fourteen miles long in the mountains of Nelson – was completed in 1862. The Public Works Department favoured a small 3 foot 6 inch gauge, with light track weighing only forty pounds per yard; and this decision severely restricted the choice of suitable locomotives.

The dearth of navigable rivers in Australia added to the importance of railways. Yet even there, water played a positive part – as it did in most parts of the world – in the development of steam transport. The fact that there was a regular steamship service between Melbourne and Sydney from the 1840s onwards made it less important that there should be rail communication between the two cities.

In the United States, where there was no shortage of great rivers and where ocean travel – rough, hazardous and for centuries slow – was an indispensible link with the outside world, it is not surprising that American inventors had turned first to steam navigation rather than steam locomotion, or that the golden age of the steamboat preceded the age of the railway.

Like the railway, steam navigation had a long pre-history. We can go back, if we choose, to a far earlier technology in China, where paddle

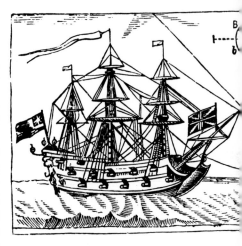

Jonathan Hulls' steam tow-boat was patented in 1736, long before a successful rotative engine had been introduced. His method of achieving rotative motion, illustrated in the inset above, involved the complicated use of ropes, grooved wheels and a counterpoise weight. Although Hulls claimed that it was a practical system, he was never able to demonstrate it as such.

wheels were first used with manpower in the 8th century. Most accounts of the origins of the steamship development begin in 1543, when a Spaniard, Blasco de Garay, was reputed to have moved a boat by steam in the harbour at Barcelona. (Spectators were not permitted to inspect the apparatus, one part of which was a vessel of boiling water.) Papin in 1690 proposed to use a piston engine driving paddle wheels to propel vessels. In the early 18th century, when economic incentives chiefly favoured the development of steam power in mining, the idea of steam navigation was nevertheless actively pursued. A proposal to use Newcomen engines for propelling a boat by forcing water out backwards figured in an English patent taken out by John Allen in 1729, and seven years later Jonathan Hulls of Gloucester took out a patent for a steamboat, accompanied by a sketch showing a steam tugboat pulling a ship. The engine Hulls had in mind was doubtless a Newcomen-type atmospheric engine, and the driving force was to be transmitted to the axle of a paddle wheel by friction mechanisms. Although the engines of his day were too heavy to meet his (or Allen's) specifications, Hulls wrote a pamphlet a year later urging that his scheme was 'practicable' and 'if encouraged' would be 'useful'.

Interest in the subject was not confined to England. A prize was offered by the French Academy in 1753 for the best essay on the manner of impelling vessels without wind; it was won by Daniel Bernouilli, who suggested a device similar to a propeller, to be operated not by Newcomen engines but by men or horses. By contrast, the Abbé Gauthier suggested a system of using engines to propel paddle wheels, for which heavy Newcomen engines were equally inappropriate. Twenty years later, Count Joseph d'Auxiron, who may have been aware of Watt's work, was granted a monopoly of steam navigation on rivers provided that he could prove his plans practicable. His attempts failed – as did those of J-C. Périer, the best-known engineer in France – although success in driving a small steamer, the *Pyroscaphe*, was reported at Lyons in 1783 by d'Auxiron's successor, Claude Jouffroy d'Abbans.

American experiments in steam navigation go back before independence. William Henry of the village of Lancaster, Pennsylvania, visited England in 1760, where he learnt of work on the steam engine, and in 1763 he constructed a steam engine which he placed in a boat fitted with paddle wheels. This and later attempts of his failed, but Henry had no doubt that 'such a Boat *will* come into use and navigate on the waters of the Ohio and Mississippi'.

After independence, John Fitch, an ingenious Connecticut mechanic, tried out a boat with paddle wheels in 1785. In the same year, he presented a model to the American Philosophical Society at Philadelphia, and within the next two years he secured patents in New Jersey, Pennsylvania, Delaware and New York. He finally carried out public trials in 1787. Most Americans were suspicious of his 'smoky' boat that 'carried fire', and his patent claims were disputed by James Rumsey, who is said to have propelled a boat by steam on the Potomac in 1786 at the rate of four miles an hour in the presence of General Washington. Rumsey's boat was driven by what we would now call hydraulic or jet propulsion. His work, described in a treatise of 1787 'on the Application of Steam', was sufficiently highly regarded for a Rumsey Society to be set up in Philadelphia, and in 1839, nearly fifty years after his death, his achievement in 'giving to the world the benefit of the steam boat' was commemorated by a gold medal presented to his son. Fitch was less esteemed. After an unsuccessful visit to France, where he obtained a patent in 1791, he returned a ruined man and attempted to commit suicide. 'The day will come,' he told his contemporaries, 'when some more forceful man will get fame and riches for my invention.'

A third American, Robert Fulton, is most usually associated with the invention of the steamboat. He had begun experimenting with paddle-wheel boats as a boy in 1779, and after a long stay in England and France proposed plans for steam vessels both to the United States and the British governments in 1793 and to the French government ten years later. In 1803 he first produced a model of a side-wheel boat and soon afterwards a working vessel, which was tried out successfully on the Seine in the presence of a huge crowd, including members of the French Academy. The water-tube boiler of this vessel is preserved in the Conservatoire des Arts et Métiers in Paris.

When Fulton returned to the United States from Europe in 1805 after twenty years in creative exile, he knew as much about steam as any of his contemporaries. He had been in contact with Boulton and Watt as well as with French engineers and had seen in action on the River Clyde in Scotland what has been described as 'the first practical steamboat', the *Charlotte Dundas*, devised by William Symington in 1801. It employed a Watt double-acting engine turning a crank on a paddle-wheel shaft. Fulton employed a Watt-type engine also when on his return to the United States he constructed the now famous *Clermont*, which from 1807 onwards operated pleasure trips on the Hudson River between New York and Albany. A contemporary called her 'a monster moving on the waters, defying wind and tide, and breathing flames and smoke'; burning not coal but pine wood for fuel, she moved further than any steam boat before, attaining a speed of five miles an hour.

Fulton built many other successful steamboats, including one 'steam vessel-of-war' ordered during the war against the British in 1812: named *Fulton the First*, it was launched in 1814. Meanwhile, in enemy Glasgow, Henry Bell had built his famous steamer *Comet*, driven by a vertical single-cylinder engine of three horse-power. She had paddle wheels on each side, and carried regular passengers between Glasgow, Greenock and Helensburgh. The company that owned her charged four shillings for the best cabin and three shillings for the second. She was described as 'a handsome vessel to ply upon the River Clyde . . . to sail by the power of air, wind and steam'. By 1814 five Scottish steamers

Overleaf:
Two paddle-steamers go all out in a midnight race on the Mississippi. This lithograph was one of 7,000 prints published between 1854 and 1880 by Nathaniel Currier and James Merritt Ives, who became almost the official portraitists of 19th-century America.

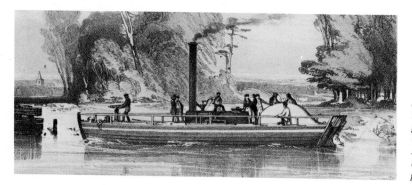

Steered by a wheel at the bows, the Charlotte Dundas *proves her worthiness to a group of gentlemen passengers on the Forth and Clyde Canal near Glasgow. She had a double-acting, single-cylinder engine, installed by William Symington in 1801, which drove a single paddle wheel in a boxed-in recess at the stern.*

In a suburb of Tokyo, a train passes along the line to Yokohama, opened in 1872 and described in this Japanese wood engraving as a tourist attraction.

'The Lackawanna Valley' by George Inness harmoniously blends human, natural and industrial themes as a boy on a grassy bank watches a puffing coal train.

'A View in White Chapel Road 1830' was a prophecy by Henry Alken of air pollution, traffic congestion and social disruption to come. Set free by steam locomotion, ne'er-do-wells would rush in from Essex and beyond, introducing wanton habits to the London streets.

Wallis's Locomotive Game of Railroad Adventures took players to many stations, ports and distant places, through tunnels and over bridges, with accidents added along the route for realism. The players drew numbers and moved counters forward until one reached 49 and won the game.

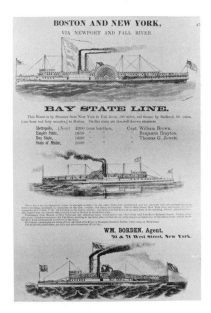

BOSTON AND NEW YORK,
VIA NEWPORT AND FALL RIVER.

BAY STATE LINE.

WM. BORDEN, Agent,
70 & 71 West Street, New York.

An advertisement for paddle-steamer services along the American East Coast stresses the quality of the ships, 'substantially fitted up with everything calculated to contribute to the ease, comfort and safety of travellers', the regularity of the sailings, and the convenience of the rail links at the ports.

were regularly working in British waters, and by 1820 the number had risen to 34, half of them in England.

The first steamer to make a sea voyage was Colonel John Stevens's *Phoenix*, which in 1809 steamed the thirteen miles from Hoboken to Philadelphia, and very soon men on both sides of the Atlantic were dreaming of crossing it by steam. The Americans won the race when in 1819 the steamer *Savannah* from Georgia travelled to St. Petersburg via Great Britain and the North European ports. It was a testing trip: all the coal on board had been consumed before the *Savannah* reached Ireland, and the return trip to the United States was carried out under sail alone. The first steamship to make the journey west was the British steam warship *Rising Star*, built in London in 1821; she had internal paddle wheels and crossed from Gravesend to Valparaiso. It was another British ship, the *Sirius*, however, which in 1838 was the first vessel to complete a transatlantic voyage entirely by steam. These early journeys to what were seen as romantic destinations quickly popularized steam navigation, influencing among others Sadi Carnot. When reflecting on the 'motive power of fire' in 1824 (see above, p.93), he selected steam navigation as one of his major social themes: it would 'bring together the most distant nations' and 'unite the nations of the earth as inhabitants of one country'.

In fact, steam was to be used, as Fulton had already used it, for purposes of war as well as of peace, and it was to serve more as an instrument of national integration, especially in countries with great distances, than of international interdependence.

As early as 1811, the first steamboat was launched west of the mountains in the United States – on the Ohio River – and the subsequent opening up of the Western frontiers would have been impossible without steamboats, small, light, fast and inexpensive. The zenith of the life of the Mississippi River, 2,350 miles long, came with the steamboat, but rivers farther west were transformed too as the steamboat arrived. Thus, along the distant Colorado River, large numbers of paddle-wheel steamboats continued to provide the cheapest and most efficient form of transport in the West for more than fifty years after the Californian gold rush of 1849. At first, the development of the railway intensified the demand for connecting steam transport by water: only at the end of the century was the balance changing.

There are many fascinating contemporary prints of intrepid steamers like the *Explorer* venturing through the Colorado canyons and of the large, elaborate baroque Mississippi pleasure boats such as the famous *Delta Queen*. There was also a rich literature, for writers were drawn irresistibly to 'the hypnotic swish of the paddlewheel'. Fitch had asked to be buried on the shore of the Ohio that he might lie 'where the song of the boatman would enliven the stillness of his resting place, and the music of the steam engine soothe his spirit'; and for many writers on river life there seemed to be genuine music in steam. No writer on railways caught the infectious spirit of invention as fiercely as Mark Twain, who was a pilot on the Mississippi from 1857 until the Civil War, and no book captures experience of river travels so vividly as his *Life on the Mississippi* (1883).

The great river boats offered unrivalled opportunities for escape, for pleasure and for excitement. They used high-pressure engines, and during the pioneering years they were dangerous on account both of their technology – with a full head of steam their boilers often exploded – and of the difficulties involved in navigating rivers that were often shallow and obstructed by sandbanks and concealed rocks. (Some captains are said to have joked that they could run on a heavy dew.) Their motto was 'Go ahead anyhow', and they usually did, often in competitive races which, while they added to the excitement, also added to the danger.

Steam did not inspire quite the same sense of exhilaration when it began to be used on the rivers and lakes of Europe. 'The finest steamboats we have in Europe are much inferior to the smallest, the wretchedest ferry boat here,' wrote an Italian Count visiting the United States near the beginning of the century – an early example of praise of American technological superiority. Nevertheless, by 1850 large numbers of steamers plied the Rhine, the Danube, the Swiss Lakes and other navigable waterways in Europe; in the coastal and ferry trade, steamers were soon carrying tens of thousands of passengers. By 1822 the *James Watt* was running a service between London and Leith, and three years later a total of 50,000 people were said to have travelled on a new steamer service from London to Margate.

Belgium, which led with boats as with railways in planning and organisation, had a *Societe anversoise des bateaux a vapeur* as early as 1835. France had 229 steamers by 1842 and 364 by 1882, some of them travelling across the Mediterranean to Italy, Spain and the Levant.

In the ocean trade Great Britain was ahead in record-breaking exploits, especially on the transatlantic route. Steamers drastically cut down travel times. Clippers had taken thirty-five days to travel from Le Havre to New York and twenty-five days with favourable winds from New York to Le Havre. The *Sirius* crossed the Atlantic in eighteen days ten hours. Only a few hours later the *Great Western*, especially constructed for the North American passenger service and lavishly equipped, arrived after a crossing (from Bristol) of fifteen days fifteen hours. The *Sirius* returned by sail; the *Great Western* made sixty-four voyages by steam alone between 1838 and 1847, and went on to serve in the West Indies for another ten years.

The *Sirius* was one of the first steamships to be fitted with a surface condenser, patented in 1834 by Samuel Hall, so that fresh water could be used and re-used in the boilers in place of salt water. Soon important technological developments with engines took place, and iron rather than wood hulls came into use. Paddle wheels could be fragile and vulnerable at sea, and the general shift from sail to steam had to wait until the introduction of the screw propeller, an idea discussed in more than one country from 1729 onwards. John Stevens was an important figure in the story, although Joseph Bramah, Watt's critic, had patented, but not used, a screw propeller in 1789 and Josef Ressel worked on the device too. The inventor who did most to ensure its adoption was a Swede living in England, John Ericsson, who secured a patent in 1836. Francis Pettit Smith, an Englishman who experimented with propellers on the Paddington Canal, took out another patent in the same year.

A symbol of progress was Brunel's *Great Britain*, an iron ship fitted with screws which left Bristol in 1843. Named and launched by the Prince Consort, who travelled from Paddington to Bristol in a special

Isambard Kingdom Brunel's Great Eastern *stands ready for launching at Millwall on the Thames (above) in November 1857. At 18,915 tons, she was more than six times bigger than any ship afloat. The first launch attempt failed, and it took three months to get her into the water. The rest of her career was similarly chequered. In New York (right), she attracted only 191 passengers when she was chartered to take American visitors to the Paris Exposition of 1867, though she had berths for 2,986. Unlucky, underpowered, unmanageable in bad weather, she nevertheless remained unsurpassed in size for nearly 50 years.*

train accompanied by Brunel and Gooch, the *Great Britain* was one of the line of famous Brunel ships. She crossed the Atlantic in fourteen days twenty-one hours, though on her way back she broke a propeller. She was followed in 1854 by the 18,915-ton *Great Eastern*, a paddle- and propeller-driven ship. This 'great Leviathan' got the maximum amount of publicity in the newspaper press, just freed from the last stamp duties – first good publicity, then the worst of all publicity as difficulties in finishing her and launching her increased. She never reached her design speed of eighteen knots. Her remarkable history, including her part in laying the transatlantic and other telegraph cables, ended in 1888

when she was sold piecemeal by auction. She had been built far too large for the technology of her time.

The improvement of that technology was one of the outstanding achievements of the age of steam, although progress from paddle wheel to propeller was slow, and accompanied by what Thurston, writing in the 1870s, called 'engineering blunders and accidents that invariably take place during such periods of transition'. Heavy, long-stroke, low-speed engines gave way to lighter engines with small cylinders and high piston speeds as the screw propeller was introduced.

.As early as 1862, John Elder secured a patent for triple- and quadruple-expansion engines; nine years later the first triple-expansion engine in the world was fitted in a French vessel; and thirteen years later the first quadruple-expansion engine was introduced. The great ocean liners of the late 19th century represented the climax of this development, for in the new century the steam turbine (described in the next chapter) was to supersede the reciprocating steam engine at sea. In 1904 the Cunard Line made the important commercial decision to instal turbines in their new ships the *Lusitania* and *Mauretania*.

Samuel Cunard, born in Canada in 1787, had established the British and North American Royal Mail Steam Packet Company in 1839 with a fleet of four 'sister ships', the first of which was the *Britannia*. Charles Dickens travelled to the United States on her in 1842, and reported that he had had an uncomfortable trip. The smoking room was only a compartment over the boiler, shared with the stokers, and the only walking space was on top of a deckhouse. Even at best the passenger standards of that time were derived from sailing-ship days, when passengers were expected to provide their own bedding and furniture.

The Americans, like Brunel, aimed to increase comfort, and they introduced healthy competition. The American Collins Line, the first of the competitors, went so far as to carpet cabins and to use steam heating on its new ship the *Atlantic*, built in New York in 1849. But when the subsidy which the company received from the state was withdrawn in 1858 the line collapsed.

New companies turned to steam on other oceanic routes. The first trip by steamer from Europe to Calcutta was made in 1825 by the *Enterprise*: her 103-day journey was fraught with almost as much excitement as a journey of the 20th-century imaginary American space-ship with the same name. Yet most ships on the Indian, Chinese and Australian routes continued to use both sail and steam after the Atlantic ships had been converted to steam alone. A landmark date was 1840, when the Peninsular Steam Navigation Company changed its name to the now familiar Peninsular and Oriental Steam Navigation Company, the P. and O.; two years later the company inaugurated a new service by paddle steamer, the *Hindoostan*, from Suez to Madras and Calcutta. The main incentive for introducing steam on such routes was to secure lucrative mail contracts, which guaranteed substantial financial rewards and demanded a constant effort in order to maintain the efficiency stipulated in the contract terms.

The growth of steam navigation coincided with the great movements of emigration by sea, which Benjamin Disraeli compared with the movements of the barbarians in the last years of the Roman Empire. The United States, the land of liberty, 'the last best hope of earth',

On canals and rivers, steam often took modest forms. The paddle-steamer Totnes Castle *(below), coming in to a landing stage on the River Dart in Devon, carried small parties on pleasure cruises – in this case less than a dozen people. At Buckby locks on the Grand Union Canal (bottom), the steam boat moored on the left is almost indistinguishable from the horse-drawn barges with chimneys for cooking fires: it has a funnel as well. Moving with the quietness of steam, such craft blended easily with the traditional waterborne environment.*

beckoned both the ambitious and the oppressed. Australia and New Zealand also offered new hopes. The first comfortable paddle-steamer to make the trip to the Antipodes in 1852–53, during the Australian gold rush, was called the *Golden Age*.

During the 1820s, only about half a million people emigrated to the United States, though this figure was more than the whole population of Australia in 1850. During the 1830s and 1840s, however, there were two and a half million emigrants to America, and during the 1850s still more – 2.7 million. During the great famine of the 1840s the Irish were preponderant amongst the migrants, and contributed much to the texture of American history. Indeed, one American writer claimed that 'of the several sorts of power working at the fabric of the Republic – water-power, steam-power and Irish-power', none worked harder than Irish-power.

Mercantile tonnage increased as dramatically during the 1840s as the emigration figures, with Britain well in the lead (she had more tonnage than the United States, her nearest competitor, France, Norway, Denmark, Holland, Hamburg and Bremen put together); Liverpool replaced Bristol as the major port. Yet the proportion of steam-driven vessels in the tonnage of the national fleets, commercial and naval, remained low even during the mid 1860s, when the British Register included 901,000 steamship tons and 4,937,000 sailing tons, the latter a peak figure in British history. The 1850s was the decade of 'the last and greatest days of the British sailing ship', and it was not certain that steam would supplant sail until the 1870s and 1880s. By then there were new opportunities for moving freight, following the development of re-frigeration. The inventors of refrigerators used gas and air, not steam, in their machinery, but the frozen products travelled by steamship. The first consignment of frozen meat arrived in London from America in 1874, with New Zealand lamb following a few years later. It became highly profitable during the 1880s and 1890s to move meat and fruit great distances to rapidly developing new urban markets.

The last chapter in the story of locomotion was the application of the steam engine to travel by air, the dream of Erasmus Darwin. An annual with the title the *Year Book of Facts* included a reference in 1844 to an 'aerial transit machine' invented by Charles Hanson – 'a car in which not only are the passengers to be placed, but likewise a steam engine, and the quantity of fuel necessary for . . . a voyage'. Hanson's machine was one of many unrealized steam fantasies; it never flew. Fourteen years later the same annual described 'a new aerial ship', the *City of New York*, a balloon fitted with a 'metallic life-boat', in which was placed an Ericsson engine and propeller. 'The sanguine projector expected,' according to the *New York Times* report, that he would 'deliver a copy of Monday's [*New York*] *Times* in London on the following Wednesday.' The *New York Times* added that 'the precise time for the first ascension has not been fixed'.

However, in 1890 a steam-powered monoplane built by a French in-ventor, Clement Ader, flew for 165 feet at a height of about eight inches. Four years later Hiram Maxim, the American-born British inventor of the Maxim machine-gun, made a similarly short hop in a biplane power-ed by a two-cylinder compound steam engine. For these brief instants the history of steam power and the history of aviation overlapped.

A crowd watches from the quay as the Queen
Elizabeth, *newly fitted out after troopship duty
during the war, leaves Southampton for New
York in 1946. The grand days of the ocean
liner were back again, reviving memories of the
1920s when dancers in evening dress created
shipboard tableaux for society photographers
(above).*

QUEENS OF THE SEA

In overseas passenger travel the final triumph of steam was the turbine-driven liner. With turbines ships could go faster and yet get bigger. The 25-knot *Berengaria*, completed in 1912, was 52,000 tons; the 32-knot *Queen Elizabeth*, launched in 1939, was 83,673 tons. Carrying thousands of passengers and sumptuously appointed, such liners created a way of life not seen before or since.

140/

An engineer officer inspects the Queen Elizabeth's *starboard propeller during a routine overhaul.*

A turbine cover is swung aboard the liner Aquitania, *nearing completion on the Clyde in 1913.*

Workmen scale and repaint the 16-ton anchors and 2-cwt. chain links of the Queen Elizabeth. *More than 7 tons of paint were put on the 1,031-foot hull.*

Polish and Czech immigrants struggle aboard ship for America in 1921, and a laden passenger (inset) joins the Queen Mary *on her maiden voyage in 1934.*

A young member of the watch takes over the wheel of the Carinthia *on a transatlantic voyage in 1930.*

Bellhops hold out their hands for inspection at a parade held twice daily, before lunch and dinner.

On the P. & O. liner Carthage, *chefs put the finishing touches to a Christmas dinner for 300 passengers.*

Passengers recline in the Social Hall of the United States Lines' Leviathan *in 1923. This and other photographs were found in a secret panel marked X.*

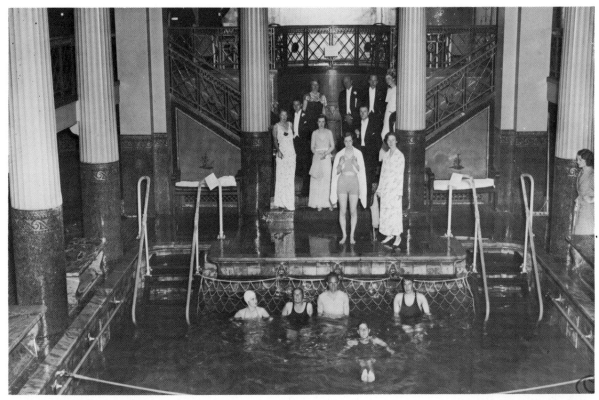

Swimmers and onlookers enjoy the pool of the White Star liner Majestic, *the former* Bismarck, *taken from the Germans after the end of the First World War.*

Waiters prepare for a meal in the first-class dining saloon of the Homeric, *another ex-German liner. It had a reputation for steadiness in rough seas.*

Battling for Passengers

Before 1914 millions of emigrants had travelled steerage to America. In 1924 the United States closed its doors, and the shipping lines introduced a 'tourist third' class to attract new low-price business.

A steward serves first-class passengers on deck.

Less well-off passengers dance to an accordion on Canadian Pacific's Duchess of Bedford *in 1931.*

Tourist-third passengers relax on deck in mid-Atlantic. They paid around $150 return, compared with the cheapest first-class returns of at least $1,000.

On the upper deck of the Empress of Britain, *players enjoy a game of tennis. Such pleasures were doomed by the coming of jet airliners in the 1950s.*

POWER FOR THE WORLD

And last of all, with inimitable power and with 'whirlwind sound', comes the potent agency of steam. In comparison with the past, what centuries of improvement has this single agent comprised in the short compass of fifty years. . . . Steam is found in triumphant operation on the seas . . . on the rivers . . . on the highways . . . at the bottom of mines . . . in the mills . . . and in the workshops. It rows, it pumps, it excavates, it carries, it draws, it lifts, it harnesses, it spins, it weaves, it prints. . . . What fuller improvement may still be made in the era of this astonishing power it is impossible to know, and it were vain to conjecture.

DANIEL WEBSTER, quoted in R. H. Thurston, *A History of the Growth of the Steam Engine* (1872)

A Fowler engine with an Indian crew, hauling cotton through the streets of Bombay, perfectly represents the sources of that city's greatness. Trade in cotton from Gujerat and steam power, applied to Indian-owned cotton mills from the 1850s on, made Bombay the Manchester of the East.

Making steam work in the mid- and late-19th century after the development of locomotion was a far more varied business than it had been in the 18th century. Although statistics for the use of steam power are patchy and limited – and some guesses have been as wild as Trevithick's fancies – there is no doubt about the basic fact that steam engines of many different kinds were being applied in large numbers of very different industries, and in agriculture, too, in an increasingly wide range of countries.

Among the new applications of steam, printing received much attention, largely because the printing press was singled out by enthusiastic Victorians as the first of the great inventions which had transformed the world. At the time of the Rainhill locomotive trials in 1829, a journalist had compared 'the impulse to civilization' given by the victory of Stephenson's *Rocket* to that of the Press which 'first opened the gates of knowledge to the human species at large'; railways and telegraph were to transform news gathering in the next twenty years. Already *The Times* was being printed by steam power. A Saxon printer living in London, Frederick Koenig, had invented in 1814 a press in which the typebed was propelled backwards and forwards by steam power; the paper was fed round a cylinder pressing on the top of the typebed. Although Koenig was not the first man to patent a steam-powered press (an earlier patent had been registered in 1790), he was the first to construct one. The invention was taken up at once by John Walter, owner of *The Times*. Hitherto the only power available to the printer had been muscle power, which even with great effort limited the number of impressions to about 250 an hour; the new machine would secure 1,000 an hour. The change, Walter claimed, was 'the greatest improvement connected with printing since the discovery of the art itself'. He had to introduce the new presses secretly, however, for fear of machine-breaking by printers worried that the new technology would put their jobs at risk.

By the middle of the 19th century most of the great London printers employed steam engines, and the great expansion of printing removed most fears of technological unemployment. Further inventions, particularly a switch from reciprocating to rotary motion, greatly increased speed and output. The American firm of R. Hoe and Company were pioneers in this field, and provided machines for the *Manchester Guardian* in 1858. Steam costs became a significant item in newspaper budgeting.

The aesthetics of steam devices continued to appeal. When James

Three American-built Avery steam tractors line up across the Canadian wheatlands around 1912. Their high-standing appearance was caused by Avery's unusual habit of placing the cylinders, flywheel and gear train under instead of over the boiler. The three engines, each developing 30 brake horse-power and pulling ten Cockshutt ploughs, broke a strip 35 feet wide. Travelling 18 miles, they thus ploughed 56 acres a day.

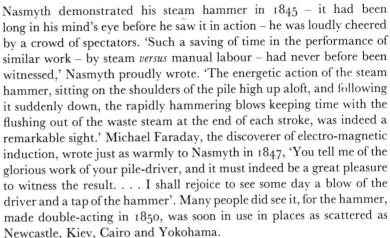

Nasmyth demonstrated his steam hammer in 1845 – it had been long in his mind's eye before he saw it in action – he was loudly cheered by a crowd of spectators. 'Such a saving of time in the performance of similar work – by steam *versus* manual labour – had never before been witnessed,' Nasmyth proudly wrote. 'The energetic action of the steam hammer, sitting on the shoulders of the pile high up aloft, and following it suddenly down, the rapidly hammering blows keeping time with the flushing out of the waste steam at the end of each stroke, was indeed a remarkable sight.' Michael Faraday, the discoverer of electro-magnetic induction, wrote just as warmly to Nasmyth in 1847, 'You tell me of the glorious work of your pile-driver, and it must indeed be a great pleasure to witness the result. . . . I shall rejoice to see some day a blow of the driver and a tap of the hammer'. Many people did see it, for the hammer, made double-acting in 1850, was soon in use in places as scattered as Newcastle, Kiev, Cairo and Yokohama.

By taking the human toil out of forging wrought iron and steel to form engine and machine parts, the steam hammer fulfilled a need. Without the existence of such needs neither the aesthetics of steam nor its 'morals' – the principles behind the gospel – could have been canvassed effectively. The potential and usefulness of steam grew enormously as greater understanding of the physics of steam power and the introduction of precision machine tools brought advances in engine design and construction. Nasmyth had worked in the great London machine-shop of Henry Maudslay, which perhaps more than any other workshop influenced the improvement of mechanical practice. The influence soon spread to the north, where already by the 1830s there were many great workshops like Matthew Murray's Round Foundry in Leeds. Nasmyth established his own Bridgewater Foundry at Patricroft between Liverpool and Manchester, and in Manchester another pupil of Maudslay's, Joseph Whitworth, introduced precision tools capable of working to one hundred thousandth of an inch.

In mid-century the rotative beam engine of Boulton and Watt began to give way to more modern horizontal engines and other new types. The idea of lying a cylinder on its side and coupling the piston rod to a crank without the intermediary of the beam was Richard Trevithick's, but the old design died hard and it was not until the 1840s and 1850s that engineers, led by Matthew Murray of Leeds, started to produce horizontal engines in any numbers. Much cheaper to build, instal and house than the beam engine, the simple horizontal engine emancipated the steam engine, transforming it into a versatile source of power that could be tailored to any kind of machinery. Engines appeared with sloping cylinders, oscillating cylinders and finally inverted cylinders, as in Nasmyth's steam hammer, and higher steam pressures enabled more and more power to be obtained from smaller cylinders.

In charting the diffusion of steam power, however, the technology can never be separated from the economics: indeed, technical history divorced from economic history can mislead. As engine types multiplied purchasers had to compare relative installation costs, running costs for both fuel and labour, and also the costs of alternative forms of power, old and new. Technical enthusiasts who failed to make the right calculations could soon find themselves in financial trouble.

In Britain steam power already exceeded water power in total power

At his works near Manchester, James Nasmyth takes the controls for a demonstration of his steam hammer in about 1851. The hammer made possible the forging of iron beams and plates larger than ever before, but could also descend so lightly that it would merely crack an egg shell.

resources by the mid 1830s, when the use of wind power was at its peak; although water power continued to increase gradually until 1870, steam power grew by sixty per cent every decade between 1840 and 1870 – a faster rate of growth than during the first thirty years of the 19th century or during the whole of the 18th century. By 1870 total horse-power in use in Britain was around two million. Factory returns (incomplete, because the Factory Acts did not cover all industry) and other supplementary evidence suggest that steam was by then preponderant not only in textiles but in most other factory industries and in most heavy industries, including iron, steel and other metals. When the first comprehensive census of steam power was taken in 1907, steam was found to account for more than ninety per cent of all power used – then over ten million horse-power.

In other countries steam power was generally slower to take hold than in Britain. In the United States, in particular, the ready availability of water power made the use of stationary steam engines initially unattractive. In New England, which was one of the first centres of industrialization, the textiles industry still derived less than thirty per cent of its power from steam. The great textile manufacturing centre of Lowell, which prided itself on its industrial paternalism and was visited by traders from all over the world, relied on water (at first using waterwheels devised on Smeaton's principles). The power of steam was not fully felt in the United States until after the end of the American Civil War in 1865, when industrialization moved westwards to areas where the economic benefits of steam seemed greater and prairie agriculture became mechanized. Until then transport was the main user of steam. Steamboats accounted for almost sixty per cent of American steam-generated power in the late 1830s, according to a Congressional Report of 1839, and later in the century railways rather than textile mills inspired the American apostles of the gospel of steam. Even in shipping, steam tonnage did not exceed sail tonnage until 1893, and during the first decade of the 20th century the use of water power was still increasing in absolute terms.

The United States, however, did not lack technical know-how, and in many branches of engine building became the second big centre of steam technology. By 1838 there were 250 steam engine builders in the United States, and many of them were more advanced in the production of high-pressure, non-condensing engines which were favoured there more than in Britain. American firms were soon making engines for export to Cuba and Latin America. Steam power became an important theme, too, in American social commentary, recognised by influential writers such as Lewis Henry Morgan, who in his *Ancient Society, or Researches in the Lines of Human Progress* (1877) included 'the steam engine with its numerous dependant machines' in a list of 'the principal contributions of modern civilization'.

At the Centennial Exhibition at Philadelphia in 1876, stationary steam engines, steam locomotives, steam fire engines, steam farm engines, steam road rollers and steam marine engines were on show. The centrepiece of the exhibition was the famous Corliss Centennial Engine with two forty-inch, ten-foot-stroke cylinders, a steam pressure of eighty pounds per square inch, and 1,400 horse-power. George H. Corliss was a prolific inventor. In 1848 he had introduced an 'automatic gover-

Instructed by the inventor George H. Corliss (raising his hat), the President of the United States, Ulysses S. Grant, and the Emperor Dom Pedro II of Brazil turn silver-plated cranks to start the four-cylinder compound beam engine, centrepiece of the 1876 Philadelphia Centennial Exhibition. As countless machines in the vast hall were set in motion through gears, shafts and belts, the crowd broke into a great cheer.

nor-controlled drop cut-off motion', which regulated the admission of steam to a cylinder, and the following year patented cylindrical rocking valves, four to each cylinder, which enabled higher power output to be obtained in his horizontal engines. For the exhibition he offered to build a steam engine powerful enough to drive all the machinery in Machinery Hall. His Centennial Engine was the largest steam engine in the world, matchless in power and efficiency. It had a march composed in its honour, and Frederic Auguste Bartholdi, designer of the Statue of Liberty, said of it that it was so harmoniously constructed that 'it had the beauty and almost the grace of the human torso'.

By the time of the exhibition Corliss had built engines with an aggregate horse-power of five million, and for many Americans his name was identified with the age of steam as much as the names of Watt and Trevithick still were across the Atlantic. A German observer at Philadelphia was struck less by the power of Corliss' huge engine than by the fact that 'several firms exhibited steam engines in various sizes, the parts of which are mass-produced automatically on machinery so that, like the parts of American sewing machines and those of several German firms, they can be interchanged one with another'. Adaptability was becoming a goal in itself, an indispensable factor behind 19th- and 20th-century marketing.

The use of steam technology spread to Europe even more slowly than it did to the United States. To protect their headstart the British at first attempted to keep out industrial 'spies' and, until 1824, to ban the migration to Europe of skilled artisans. In spite of these efforts, knowledge of steam technology soon made its way out of Britain to France and Belgium, and plant for producing steam engines on the Continent was installed by enterprising Englishmen such as John Cockerill, founder of a metal works at Seraing, near Liège in Belgium.

The first steam engine in a French cotton mill was installed at Orleans in 1791, but during the drawn out revolutionary and Napoleonic wars the old established Chaillot Foundry concentrated on artillery; only later did it become a major producer of engines, and the use of steam power was slow to increase. Jennifer Tann and M. J. Breckin, who have collected detailed statistics of the distribution of Boulton and Watt engines in Europe, have shown that although the French made many enquiries about British steam engines, they placed few orders. To protect their own late developing engine industry, the French in 1818 imposed a thirty per cent duty on imported engines. Domestically produced engines, however, consumed large amounts of expensive fuel and were small and weak, with an average of only ten horse-power.

It has been claimed that by 1830 fifty steam engines a year were being manufactured in France, yet by the end of the 1830s English textile mills alone were employing more steam power than the whole of France did at the time of the revolution of 1848. Although by then France had more than 400,000 factory workers, the French textiles industry was still using two and a half times as much water power as steam, and half the steam engines in France were located in five of the eighty-six *départements*. Even in 1880 total steam power in France, excluding locomotives, railways and marine engines, was only 544,000 horse-power, and ten years later 803,000. The figure doubled, however, between 1890 and 1900 and reached 3,539,000 by the outbreak of the First World War in early August 1914.

Part of the explanation for this backwardness was that France was less well endowed with indigenous fuel supplies than Britain. In one respect, however, this shortcoming eventually turned out to be an advantage. From an early date the more economic high-pressure engines were preferred in France, and also in Belgium, to the low-pressure condensing engines of Boulton and Watt. Such engines were produced according to the designs of the Englishman Arthur Woolf, pioneer of the compound steam engine, whose partner emigrated to France in 1815. From this beginning France was to secure her world lead in compound-cylinder railway locomotives (*see page 111*).

In Germany, where industrialization was late and swift, occurring mainly between 1870 and 1914, many of the dramas of British economic and social history fifty or more years before were re-enacted in a new context of greater government intervention. Steam power, already advancing after 1850, came to dominate the textile industries; the number of workers employed in domestic production was drastically cut; a modern ship-building and marine engine industry was built up (steam ship tonnage increasing more than seven times between 1870 and 1890); and urbanization proceeded rapidly. In 1870 Germany had eight cities with a population of more than 100,000, in 1900 forty-one.

Not surprisingly, steam power was both rhapsodized and bitterly attacked in Germany, as it had been earlier in Britain. 'Mightier beyond compare than steed or chariot, than car or sail,' a German publisher exclaimed in 1882, 'is the new and powerful motor of our day: steam, which, with aquiline speed, guides ocean castles and rolling towns.' Dr. Ernst Engel, director of the Prussian Office of Statistics, was more fanciful. For once forsaking the 'bleak world of boiler and steam engine statistics', Engel chose the occasion of a celebration dinner in 1875, on the centenary of Watt's separate condenser patent, to proclaim the marriage of men and engines, their 'wives', as 'one of the happiest on the face of the earth'. It was also, he went on, the most fruitful. 'Its offspring are numbered in hundreds of thousands. With very few exceptions, they are the best bred, hardest working and most docile of creatures. They never rest by day or night and are veritable models of obedience and temperance. . . . Whenever we build huts for them and treat them properly, their entrance is followed by success and abundance.'

Such fancy did not impress the critics of steam. Writing decades later, Dolf Sternberger in his *Panorama of the Ninteenth Century* claimed that Engel's tone of 'smirking banter revealed the underlying demonic' nature of steam all the more frighteningly. While there was little demonic about the ploughing engines being turned out by Kemna of Breslau or Wolf of Magdeburg, it is notable that swords determined Germany's economic destiny as much as ploughshares. According to one German economist, the successful war with France in 1870–71 provided (by way of war indemnity) 'France's finishing off our main railway network for us'. Steam in this case was not a cause of peaceful progress, but a consequence of successful war.

Among the new applications of steam in the 19th century, agriculture always received much attention both in Britain and the United States and in many parts of Europe. Taking the century as a whole, it had a

An unlikely application of steam, but it existed: at a country house with plenty of grass, a sales representative opens the regulator to start a steam lawn mower with a large, view-obscuring vertical boiler. It was built around 1885 by the Lancashire Steam Motor Company, which later became Leyland Motors and, later still, British Leyland.

A mule cart sets out from a sugar mill in Barbados with a heavy load, probably of refined sugar, in 1893. Steam engines were widely employed to power sugar-cane crushing machinery during the 19th century, and when fired on bagasse waste, the cane fibres left over after crushing, they were very economical.

more marked effect on America than on Britain, and it has continued to influence developing countries since. Its history, however, began in Britain. The first farm in the world to use steam power was in North Wales – in 1798 – and was quite exceptional in that it was owned by the engineer and ironmaster John Wilkinson, Matthew Boulton's associate. A year later a steam engine was used to thresh corn in East Lothian; significantly it was located next to a coal mine, where steam machinery was already in use. An engine installed in Norfolk in 1804 was probably the first to be acquired specifically for farm work.

Nonetheless, it was another half century before steam came into common use on the farm. Mobile rather than stationary engines became the most widely used source of power. Although Richard Trevithick had pioneered the small, self-contained high-pressure engine at the very beginning of the 19th century, it took time to produce a portable engine light enough to be moved about on the farm by horses without running away or getting bogged down. The logical development of the horse-drawn portable engine was the traction engine, which could not only drive machinery but haul tackle from place to place. Several inventors experimented with self-moving engines. In 1842 J. R. and A. Ransome of Ipswich, a firm founded four years earlier and never short of ideas, demonstrated an engine at the Royal Agricultural Show at Bristol which was driven by a chain connecting the crankshaft and the wheels. Thomas Aveling of Rochester introduced simplified chain drives in 1860 and 1862, and other inventors soon began introducing gear drive, creating by 1870 the type of traction engine that was to be built for the next sixty years.

Threshing was the first agricultural task to which steam engines were applied. The poet C. P. Tennyson hailed the threshing machine in the mid-19th century as an 'augury of coming change':

'Did any star of ancient time forebode
This mighty engine, which we daily see
Accepting our full harvests, like a god?'

Steam cultivation offered the next challenge to inventors, who from 1832 onwards produced devices of every conceivable kind for ploughing, digging and 'stirring' the soil, some of them with Heath Robinson characteristics, many of them depending on the use of rails, tramlines or wooden platforms to support the engines. (Trevithick, who as always saw all the technical openings, had envisaged, though not built, what he called 'a steam spade tormentor'.) Cable-style ploughing was introduced, not very widely, during the 1830s, and in the golden age of English high farming in the middle years of the century John Fowler of Leeds (he had been born at Melksham in rural Wiltshire) was highly successful in introducing pairs of traction engines to power a rope or wire windlass by which a plough was pulled backwards and forwards across the fields. The Fowler system made it possible to plough more deeply than before without compressing the soil with heavy wheels, but the engines were expensive to buy and heavy in fuel costs. Higher steam pressures and compounding of cylinders were introduced during the 1880s, which helped to reduce boiler size and economize on fuel and so cut running costs.

The intrusion of steam into the life of the English farm was a new social experience. Generally the threshing and ploughing was done by teams of engine-men brought in seasonally by agricultural contractors, and the look of the landscape was changed as smoke rose from green fields. At least one early 19th-century English novel, *The Mummy* (1828) by J. A. Loudon, introduced the subject of steam ploughing. Loudon was a passionate steam enthusiast, but Thomas Hardy in his tragic novel *Tess of the D'Urbervilles* (1891) treated the intrusion of steam as an invasion destroying the time-old rhythm of English rural life.

In the United States steam power had been increasingly applied during the 1830s and 1840s to powering cotton gins, driving cane-sugar crushing machines and servicing rice cultivation. By the last decades of the century, when American agriculture flourished and British cereal agriculture languished, traction engines were in wide use. Ploughing in the United States, where soils were often lighter than in England, could be direct: the engines could draw the plough, because their weight did not greatly compact the soil in front of it. American engine-builders had the long-term advantage of economies of scale. J. I. Case of Wisconsin, founded in 1864, became during the next thirty years the largest producer of agricultural engines in the world. Their biggest engine was of 150 horse-power with a huge cylinder and wheels eight feet in diameter. In 1889 Daniel Best introduced a 'steam harvesting outfit' consisting of a traction engine with a 'combined harvester' attached; it was said to harvest up to a hundred acres a day in California. Best's vertical 'coffee-pot' boilers were highly distinctive, and one of his huge engines weighed sixty-five tons; it had a chimney more than twenty feet high.

Steam was applied to various other uses on the land apart from threshing and ploughing: drying grain, 'cooking' animal food (one report of 1892 described the 'forceful aroma, delicate flavour and nice, palatable condition' of steamed cattle food); steaming soil to sterilize it (as in the Guernsey tomato industry of the 20th century); and above all, drainage. A huge Dutch land reclamation scheme at Haarlem Meer in the late 1840s was carried out with three stationary engines supplied from Cornwall, the largest beam pumping engines ever built. In England the 19th-century drainage of the Fens (the latest phase in a long history) owed much to beam engines driving scoop wheels. It was reported in 1840 that the use of steam pumps in the Lincolnshire Fens had raised the value of land from £20 to £70 an acre.

At Haarlem Meer in Holland (left), three of eight beams driving large bucket pumps protrude from the circular engine house of the Cruquius pumping station, built in 1849 to drain the meer and now preserved. All eight beams were operated by one annular compound engine.

Inside Papplewick Pumping Station (below), a waterworks near Nottingham, the pump rods, connecting rod and 24-ton flywheel of an 1884 Watt rotative beam engine gleam in preservation.

On both sides of the Atlantic, portable and traction engines powered equipment in fairgrounds as well as in the fields. They drove swings and roundabouts, generated electric light, and hauled the showman's caravans and trucks from site to site. They were the most highly decorated emblems of the age of steam, except perhaps for the great river boats in America. Steam was first applied to a roundabout in about 1870 by Henry Soame of Mersham, Norfolk, in the heart of a farming district; he took the power from a portable engine by means of a flat belt drive. (The word roundabout had earlier been used for systems of ploughing.) Frederick Savage of Kings Lynn became the leading manufacturer of portable engines adapted to driving fairground machinery – a spin-off from his agricultural machinery business. Steam engines also powered the bellows and mechanisms of fairground organs, and the American 'calliope' used steam directly to create raucous fairground music. It was a portable boiler carrying an array of steam-blown whistles. Tunes could be played on it as it was towed along, usually by horses, and it was often used to herald the arrival of a circus. Whistling steam fire engines could also be heard as well as seen: some were steam propelled, particularly in the United States.

The numerous progeny of the traction engine also included steam rollers. A Frenchman, Louis Lemoine of Bordeaux, patented the first in 1859. Thomas Aveling produced a twenty-two-ton machine in 1866; it spent two years rolling the roads in Hyde Park in London, working only at night so as not to frighten the horses. Within a few years, British-built steam rollers, designed round the basic traction engine with a single roller in place of the front wheels and two smooth wheels at the rear, were being exported to many countries including the United States, where the tandem roller (with two equal-size rollers back and front) was introduced. Without the steam roller, roads could never have been made suitable for high-speed mechanically propelled vehicles. Indeed, the poor condition of the roads was one of the reasons why passenger and freight transport continued to depend on the railway for most of the 19th century. Another reason, in Britain especially, was the opposition to steam vehicles from conservative farmers, irritated villagers and selfish toll companies which charged punitive rates for the use of their roads.

Road steamers had a long pedigree that stretched back before Trevithick to Cugnot and his experiments in the 1760s and 1770s. Indeed, steam carriages were seen regularly on the roads before traction engines came into general use on farms. Walter Hancock had patented a boiler for a road engine in 1827; his *Infant*, one of nine steam carriages he built, ran to Brighton from London in 1832, carrying a party of eleven people at a speed of nine miles an hour. Another early steam carriage invented in 1833 by Michael Roberts of Manchester embodied the first application of a differential gear to a vehicle while a third was described in *The Lady's Magazine* in 1828 as 'preserving good speed in ascending hills' and descending 'under perfect control'. Farey claimed that steam engines were safer than carriages drawn by horses, and far more manageable.

Nevertheless, the anti-steam lobby grew strong on public complaints. During the 1830s Hancock's steam carriages aroused animosity on the grounds that they created unpleasant smoke and scared horses. A

Fire!

The steam fire engine was a late developer. The first, a horse-drawn model, was built in London in 1829. Self-propelled engines became popular in America after 1840, but even there horse-drawn types persisted, and in Britain they were standard until the motor fire engine appeared in 1904. All steam fire engines had quick-responding vertical boilers. If horse-drawn, they could get up steam on their way to a fire.

Beside the Thames, London's new Metropolitan Fire Brigade practise with steam pumps in 1868.

Pouring smoke, an American horse-drawn appliance races to a fire in 1900. With a wood fire, as here, it was possible to raise steam in 15 to 20 minutes.

quarter of a century later, opinion began to prompt legislation. Locomotive Acts in 1861 and 1865 kept down speeds in Britain by law to two
miles an hour in towns and four miles an hour in the country, nominally
on safety grounds, and stipulated that every steam vehicle using the
road had to be accompanied by three men, one walking in front and
waving a red flag. These regulations remained in force until 1896, by
which time other countries, France in particular, were well ahead in
the development of steam cars and commercial vehicles.

During the 1830s a Frenchman, Charles Dietz, had been running a
regular steam service between Paris and Versailles with vertical boiler
tractors, and Frenchmen remained prominent in technical advance on
the roads. In 1842 Jacques Meyer invented an expansion valve which
enabled more effective use to be made of the automatic speed governor.
In 1873 Amédé Bollée's steam locomotive *Obéissante* covered 138 miles
in eighteen hours, and at the end of the century Leon Serpollet's steam
cars with flash boilers were a familiar sight on the roads. By 1900 his
Paris factory was producing two hundred cars annually. One of his
1903 models, equipped with a high-pressure superheating boiler and
pneumatic tyres, reached a speed of eighty miles an hour. Strong competition in steam cars came from the United States, where a Stanley
steamer reached 127.66 miles an hour in 1906. But by that time the
cleaner, lighter and more efficient petrol engine was beginning to
take over on the roads.

By the last quarter of the 19th century, there were, as Dr. Ernst Engel
had put it, many 'offspring' of the steam engine in Germany, the United
States and other countries besides Britain, both inside and outside
Europe. Traction engines were being made in Austria, Hungary,
Poland and Sweden, and in the last of these the use of steam saws was
freeing the timber industry from dependence on coastal or river sites
with waterwheels. In Russia, where the total power of all steam engines
in use in 1840 had been as little as 20,000 horse-power, engine-making
and engine use both increased substantially after the building of railways.
In China, most 'factories' – and it was estimated that in 1912 there were
still only just over 20,000 of them – used human or animal power; only
363 employed mechanical power. Yet in Japan, willing to accept
Western ideas and techniques, the first steam hammer was used in 1859,
almost a decade before the Meiji Restoration, and 'sons of men of rank'
were described as studying a steam engine at work. The first steamer
had been completed at Nagasaki two years earlier. Many Japanese
textiles mills turned to steam, though coal was not cheap, and in 1881
five out of nineteen non-textile factories in Tokyo were using steam –
among them concerns producing paper, leather goods, matches and
umbrellas.

More steam engines were employed in colonial agriculture than in
colonial industry, but India provided an interesting example of what
could be done with colonial industrialization. During the 1850s, railway
building was associated with the growth of textile and other factories
using steam power, particularly in Bombay, Ahmedabad and the cotton
areas of Gujerat. Construction of the first Bombay cotton mill started in
1854, and six years later a Bombay newspaper boasted that the city,
which had 'long been the Liverpool of the East', had now become the

Titans of Steam

Around 1910 there was a rash of inventions designed to spread the weight of a heavy steam engine over a greater area, so enabling it to cross soft ground. The Botrail system worked, but required very cumbersome wheels. The Pedrail system was introduced to overcome this problem, but was superseded by more modern caterpillar tracks. Perhaps the oddest creation of steam was the Darby Walking Digger, though some rare dredging engines could rival it in that respect. The steamroller by contrast appeared positively conventional.

A Hornsby steam chain tractor undergoes trials.

Circular pads give a Foster engine extra grip.

A US trenching machine raises its giant arm.

A run-away steamroller teeters on the 'brink.

A Fowler Botrail Tractor spreads its weight on double-width wheels fitted with plank-like iron treads. Heavy hawsers tension the treads as they hinge.

A giant Fowler engine with tracks front and rear straddles a ditch with its winch. For a clear view, the steersman stands on a platform outside the cab.

Laden with coconut pulp, a 1928 Robey articulated steam lorry works on in Sri-Lanka in 1973.

A Darby Walking Digger, moving broadside across the field, reveals its six mechanical forks.

Manchester also: 'factory chimney-stacks . . . meet the eye on every side'. The American Civil War, by interrupting US cotton exports, stimulated development in Indian cotton. By 1901 there were 124 steam-driven Bombay factories covered by the Factories Act, and in 1910 the number had grown to 133. Considerably more than half were textile mills, but steam engines also worked in flour mills, iron or brass foundries and printing works.

The various types of engines in use in different parts of the world towards the end of the 19th century were very conveniently catalogued in terms of their mechanical design in J. A. Ewing's manual *The Steam Engine*, the first edition of which appeared in 1894. Books on steam power by then had become far more technically sophisticated and far less rhetorical than earlier in the century. Robert Willis's *Principles of Mechanism* (1841) had led the way, and W. Macquorn Rankine of Glasgow had followed with his impressive *Manual of Applied Mechanics* in 1858 and *A Manual of the Steam Engine and Other Prime Movers* a year later. Significantly both authors were professors – Willis at Cambridge and Rankine at Glasgow – as was J. A. (later Sir Alfred) Ewing, another Cambridge engineer. By 1908, even traditionalist Oxford was accepting mechanical science as a suitable subject for undergraduates.

OSCILLATING ENGINE

In general, Ewing wrote, engines were 'of the reciprocating or piston-and-cylinder type', but they could be classified in several different ways. 'There is a general distinction of *condensing* from *non-condensing* engines, with a sub-division of the condensing class into those which act by surface condensation and those which use jet condensers.' Next there was a division between *simple* and *compound* engines, with a further classification of the latter, according to the number of stages in the compounding, as double, triple or quadruple expansion engines. Again, engines were *single-acting* or *double-acting*, 'according as the steam acts on one side only or on both sides of the piston'. Some engines, such as steam hammers, were *non-rotative*, that is to say, the motion of the engine worked simply on a reciprocating piece of machinery. Most, however, worked on a continuously revolving shaft and were *rotative*.

TABLE ENGINE

'In most cases,' Ewing went on, 'the crank-pin of the revolving shaft is connected directly with the piston-rod by a connecting-rod, and the engine is then said to be *direct-acting*; in other cases, of which the *beam engine* is the most important example, a lever resembling the beam of a balance is interposed between the piston and the connecting-rod leading to the crank. The same distinction applies to *non-rotative pumping engines*, in some of which the piston acts directly on the pump rod, while in others it acts through a beam.'

'The position of the cylinder,' he added, 'is another element of description, giving *horizontal*, *vertical* and *inclined cylinder* engines. Most vertical engines have the cylinder or cylinders set above the crank-shaft, and these are sometimes described as being of the *inverted cylinder* class. In *oscillating cylinder* engines the connecting-rod is dispensed with; the piston-rod works on the crank pin, and the cylinder oscillates on trunnions to allow the piston-rod to follow the crank pin round its circular path. In *trunk* engines, the piston-rod is dispensed with; the connecting-rod extends as far as the piston, to which it is jointed, and a trunk or tubular extension of the piston, through the cylinder cover,

STEAM-LAUNCH ENGINE

INCLINED-CYLINDER WINCH

HORIZONTAL ENGINE

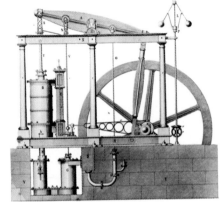

ROTATIVE BEAM ENGINE

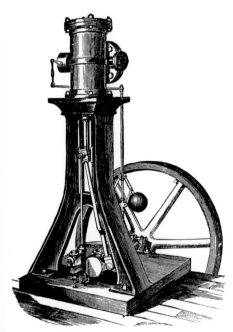

VERTICAL ENGINE

gives room for the rod to oscillate. In *rotary* engines, there is no piston in the ordinary sense: the steam does work on a revolving piece, and the necessity is thus avoided of afterwards converting reciprocating into rotary motion.'

Ewing still had not packed all his distinctions into this one, rather breathless paragraph. 'Still another mode of classification,' he went on, 'speaks of engines in reference to the conditions under which they are at work as *stationary, locomotive* or *marine*. Locomotive, marine and some kinds of stationary engines such as those employed in heavy rolling mills belong to the *reversing* class, having valve mechanism which enables them to run either way'.

Ewing's classification had nothing to do with chronology and paid little attention to function. These aspects of the history of steam power are of more interest, however, both to historians and to connoisseurs. As far as chronology is concerned, the beam engine, of course, had come first, starting with Newcomen's single-acting atmospheric engine, which used steam as a convenient method of creating the vacuum from which it derived its power, and continuing through the age of Watt. As far as function was concerned, the beam engine (with cast iron or later wrought iron beams replacing wood) was appropriate for pumping, the first of the purposes to which steam had been applied, and for many other purposes besides. A direct but much more efficient derivative of the Watt pumping engine was the 19th-century 'Cornish engine', which used high-pressure steam and went through a long process of refinement; the last Cornish engine, used to drain the Severn Tunnel, did not cease work until 1962.

Some single-cylinder beam engines were converted into compounds, following the course pioneered by Jonathan Hornblower, Trevithick and Woolf, thereby significantly reducing coal consumption. A key figure in the story of compounding was John McNaught of Bury, who in 1845 added a high-pressure cylinder to a Watt beam engine, placing it on the opposite side of the beam from the existing low-pressure cylinder.

The word 'to McNaught' now entered the language. Engine layout changed as compounding established itself. Some engines had cylinders side by side, some cylinders 'in tandem', sharing a common piston rod, and some, the rarer 'annular compound' type, had the larger low-pressure cylinder enclosing the smaller high-pressure cylinder.

Murdock, Trevithick and Maudslay all experimented with engines without beams, the last of them patenting a 'table engine' in 1807 in which a vertical cylinder was mounted on a four-legged cast-iron frame or table. The beam-less type that became most popular in Britain – though not until the 1850s – was the horizontal engine. There were various reasons for its slowness to find acceptance – fear of excessive cylinder wear caused by the weight of the piston pressing downwards, for example – but perhaps the main one was the long life of many of the beam engines and the cost of replacing them. In America, where there was more willingness to treat machines as obsolescent when better alternatives appeared, the Corliss horizontal engine led the way, and there were to be further important technical developments in both the United States and continental Europe in the 20th century, including the 'super-heating' of steam in its passage between the boiler and engine and the improvement of the 'Uniflow' or 'Unaflow' engine, which stabilized the working temperature along the cylinder and thereby cut down heat loss.

Whole textbooks have been written about particular parts of the engine including valves, gears, the governor and, above all, boilers. It was the Cornish boiler which made possible the Cornish engine and ensured its longevity among the steam engine progeny. It occupied less than a third of the space required by the old haystack boilers yet provided the same heating surface. More important, it was capable of withstanding steam at the higher pressure needed if the engine were to be made more efficient. The first Cornish boilers were thirty feet long and six feet in diameter, with a central tube about three-and-a-half feet in diameter containing the fire.

Large Cornish engines with cylinders as much as eight or nine feet in diameter needed a battery of up to six Cornish boilers to provide them with enough steam; one boiler could generally be kept spare or blown down for maintenance so that the engine need never stop. In an effort to improve performance, the duty of engines in Cornwall was carefully recorded monthly in Joel Lean's *Engine Reporter*, which first appeared in 1811 and was later taken over by the Royal Cornwall Polytechnic Society. Originally used in Cornish mines, these engines made their way during the first half of the 19th century to countries as far away as Australia, Mexico and South Africa. As Cornish mining declined during the second half of the century, Cornish engines in Britain came to be used mainly in waterworks in large cities where the quantities of water required to be pumped were very great. One Cornish engine at Kew in London, now preserved, could move 3,228 litres (710 gallons) of water per stroke against a head of 175 feet; another engine in Battersea, installed in 1858, had a cylinder nine feet four inches in diameter. Such engines required huge houses, some of which were elaborately designed and constructed, though one variant of the Cornish engine, the Bull, had the cylinder over the pump well or shaft, so requiring a less complex engine house. Cornish engines were as economical in fuel as

In the worst type of steam engine accident (right), a boiler explosion on the Staten Island ferry Westfield *sends human bodies, horses and superstructure skywards at Whitehall Landing, New York, fulfilling the direst prophecies of early cartoonists. This 'appalling disaster', as a newspaper described it, occurred on 30th July 1871, costing the lives of over 60 passengers.*

Broken bearings and gear wheels (left), a governor leaning at a drunken angle and a gaping hole in the engine house mark the path of a flywheel that broke loose in a Lancashire factory around 1910. Seen by more people (below), a traction engine that ran away on a hill smokes in the front doorway of a Georgian terraced house.

they were robust in construction, and they could be used as blowing engines to provide an air blast for iron furnaces as well as for pumping.

The second type of favourite boiler was the Lancashire boiler, patented in 1845 by Sir William Fairbairn. It had two furnace tubes surrounded by a single shell. Because it had a greater heating surface, it could produce more steam than the Cornish boilers and did not require such careful control of the water level; it was to remain in existence even longer. In the United States G. A. Babcock and S. Wilcox introduced a water-tube boiler in 1867, which was safer than a shell-type Cornish or Lancashire boiler because the volume of hot water and steam in circulation was smaller for a given output.

The safety – or lack of safety – of boilers was an important part of 19th-century steam history. Boiler explosions were regrettably commonplace, especially in the first half of the century. They occurred for a variety of reasons: undetected corrosion or furring of the heated surfaces, clumsy repairs, or failure to keep the water up to the required level, so causing firebox plates to overheat. As early as 1803 Trevithick had devised the forerunner of the lead plug which was designed to melt if the firebox crown became overheated and release steam before worse damage was done. But like so many of Trevithick's ideas, it was a long time before the device came into common use.

After an explosion at Rochdale in 1854, when ten people were 'blown to atoms', boiler insurance was introduced by an association which later became the Boiler Insurance and Steam Power Company, the first of several such concerns. Not until 1882, however, was safety legislation introduced in Britain – a Boiler Explosion Act of limited scope, to be followed by a more extensive Act eight years later. In the United States there was no government regulation at all. During the debate on the first British Bill, one MP forcefully reminded his colleagues of the power of steam: 'he did not know whether Hon. Members were aware that underneath the floor of that House or very nearly so, there was a group of steam boilers working at high pressure; and if one of those boilers was to burst at the time the House was sitting, there would be a great many vacant seats in the representation of the counties and the boroughs.' As if to underline the need for the legislation, a boiler explosion took place near Rochdale, scene of the 1854 explosion, on the very day of the passing of the Act.

Following the introduction of legislation the number of lives lost in boiler explosions fell from thirty-five in 1883 to twenty-four in 1900 and fourteen in 1905: in the first of these years, a particularly bad one in the United States, the comparable American figure was 383.

The Boiler Insurance and Steam Power Company publicized what it called 'the necessity of scientific supervision'. The term was a revealing one. The age of steam had seen the emergence of self-made men with little knowledge of science. At the time of the Paris Exhibition in 1855, no less a person than Sir William Fairbairn, inventor of the Lancashire boiler, had publicly drawn attention to the problem when he attacked the rule of thumb. 'The French and Germans are in advance of us in theoretical knowledge of the principles of the higher branches of industrial art,' he argued, 'and I think this arises from the greater facilities afforded by the institutions of those countries for instruction in chemical and mechanical science.' Yet to some people the economic

A clean break in the beam of the pumping engine at Hartley Colliery, Northumberland, caused one of the worst ever steam engine accidents. When the beam snapped in 1862, the broken half plunged down the only shaft, trapping 202 men and boys who were gradually asphyxiated. Thereafter all pits were required by law to have at least two shafts.

motive rather than education or the inherent characteristics of steam technology seemed to be at fault. The owners of steam engines were often more interested in profits than techniques. As a character remarked in Thomas Love Peacock's *Gryll Grange* (published in 1860, but written before), 'high pressure steam boilers would not scatter death and destruction around them if the dishonesty of avarice did not tempt their employment'. Some owners of steam engines, an officer of the Boiler Insurance and Power Company reported, even followed such 'reckless' practices as weighting safety valve levers with broken wheels, pieces of old metal, bricks and other objects.

Already, however, there was an old-fashioned and inefficient air about such men, who belonged to what Lewis Mumford has called the palaeo-technic phase of industrial history. Mumford distinguishes this period from the neo-technic phase, drawing the dividing line in the last quarter of the 19th century. In the neo-technic phase there was a growing emphasis on precise measurement, effective technical control and scientific education as well as an interest in new metals and alloys and a revulsion against waste of resources. New forms of power were also being introduced. Gas and hydraulic power were already being used to do many things that steam engines had done (in 1846, for example, the first hydraulic crane had been invented, and in 1860 the first hydraulic forging press). Finally during the 1880s the reciprocating steam engine began to give way as a prime mover to the steam turbine.

The steam turbine was Britain's third great contribution to the age of steam, with its inventor Charles A. Parsons (sixth and youngest son of the Earl of Rosse) following in the wake of Newcomen, Watt and Trevithick as one of the major figures in the history of steam power. Parson's invention ensured that steam would continue to be used in an age of electricity. The turbine had a long pedigree, however, stretching back to Hero's aeolipyle (a simple reaction turbine) in the ancient world and to Branca's toy, a simple impulse turbine, of 1629 (*see pages* 18, 19). The modern impulse turbine, like Branca's, is merely a development of the windmill: jets of steam issue from fixed guide blades and impinge on blades fastened to a rotor, causing it to revolve. The expansion of the steam as it leaves the guide blades provides the rotative force. Interest in the turbine grew in the age of Watt, though he was as sceptical about it as he was about high-pressure steam. Trevithick devised a whirling engine, which was a version of the steam turbine, four years before Watt's death. During the late 1830s Avery in the United States invented traction steam wheels for driving circular saws and cotton gins. In Britain nearly 200 patents for gas and steam turbines were registered between 1784 and 1884, the year when Parsons took out his patent for a steam turbine and successfully constructed one.

The 1884 turbine was greeted at first as a toy rather than as a major invention of crucial importance in the development of future technology. It was not until Parsons gave an unscheduled demonstration of his 100-foot turbine-propelled boat, the *Turbinia*, during Queen Victoria's Diamond Jubilee naval review at Spithead in 1897 that the implications of the invention became dramatically plain. The *Turbinia*, streaking down the review lane at nearly thirty-five knots, made the Royal Navy's fastest destroyers seem slow: in a new and serious version of 'Catch-me-who-can', they lost out to the new intruder.

Slow or intermittent progress with steam turbines before 1884 had been due in part to the technical limitations of turbines – the fact that they were extravagant of steam and that their rotors turned too quickly. These limitations persisted until design and metallurgical and manufacturing techniques enabled engineers to produce blades closer to the correct theoretical shape and finish them to a high degree of accuracy. Because there could be no contact between the fixed and moving parts, there was always a potential steam leak, but the clearance was gradually brought down to the modern $\frac{3}{4}$mm. to 1mm. Solutions were reached in a different way from the solutions of Newcomen or Watt. They depended directly on the work of theoreticians. 'The old rule of thumb methods have been discarded,' Parsons himself wrote. 'The discoveries and data made and tabulated by physicists, chemists and metallurgists are eagerly sought by the engineer, and as far as possible utilized by him in his designs. In many of the best equipped works, also, a large amount of experimental research, directly bearing on the business, is carried on by the staff.'

Parsons was not the only inventor in the field. Carl Gustav de Laval, working independently in Sweden in 1888–89, devised a form of impulse steam turbine in which the steam was expanded through a single nozzle from a region of high pressure to one of low pressure: the turbine ran at exceptionally high speed. Eleven years later, Charles Curtis of New York invented a second type of impulse turbine. There was sharp competition between the two types. However, the high rotor speeds meant that gears had to be used between the turbine and the machine it was driving.

In 1884 Parsons, in his own words, 'dealt with the turbine problem in a different way'. It seemed to him that moderate surface velocities and speeds of rotation were essential if the turbine motor 'was to receive general acceptance as a prime mover'. In fact, Parson's first steam turbine, a long thin prototype of ten horse-power, ran at 18,000 revolutions a minute. But year by year Parsons improved his turbine. As he produced installations with larger-diameter rotors for power stations, speeds fell – to 4,800 r.p.m. in 1890, 3,000 r.p.m. in 1894 and 1,500 r.p.m in 1905 – and both power and efficiency increased dramatically. Aubrey Burstall in his *History of Mechanical Engineering* (1963) gives the following figures:

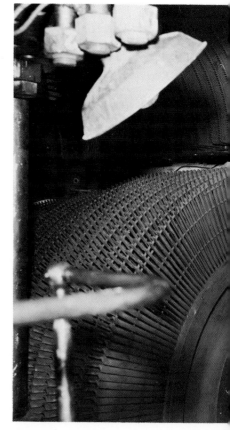

Year	1884	1892	1900	1913	1923
Power (kW)	7.5	100	1,250	25,000	50,000
Steam Consumption lbs/kWh	129	27	18.2	10.4	8.2

Patent problems arose with the steam turbine as they had done in the 18th century with the reciprocating steam engine. So, too, with the construction of a suitable dynamo to run with the turbine which, as Parsons put it, 'involved nearly as much trouble as the turbine itself'. Parsons relinquished his patent rights in 1889 on leaving the company he worked for in order to start his own, and only re-acquired them in 1893. He had them extended to 1903. There were other parallels with the 18th century too. Once again the fate of steam power was closely bound up with the fate of water power. It was not just that Parsons emphasised 'the close analogy between the laws for the flow of steam and water under small differences of pressure'. Before devising steam turbines, he was interested in the example of hydraulic turbines. So also was de Laval, whose steam turbine was much nearer to a water turbine prototype than that of Parsons. Water turbines had been developed far more in the United States and France than in Britain. Indeed, at the Great Exhibition of 1851, a French water turbine, developing 140 horse-power, had diverted attention from steam.

At the time of Parson's death in 1931, the turbine was said to have reached 'an advanced stage of development', although by 1951, the year of the Festival of Britain, a 100,000 kW electricity-generating unit had been built in London by Metropolitan-Vickers, and the largest turbine in the world was a 125,000 kW unit at Sewaren, New Jersey, in the United States. Since then world demand for power has been increasing at the rate of eight per cent per annum; steam turbines produce most of the power. In 1960 twin-shaft, cross compound turbines were introduced, and steam turbines were employed in nuclear as well as conventionally fired plant. Large modern units now produce up to 660 MW (660,000 kW).

Without electricity – in its time a revolutionary new technology – the development of the steam turbine would doubtless have been quite different. Turbines had two advantages over reciprocating engines – they took up less space and had greater fuel efficiency. Reciprocating engines, however, were widely used in the first electrical power stations before the turbine took over and continued to be used for a variety of purposes in the 20th century. In their improved form they were themselves, to use Mumford's language, neo-technic engines which could not have been manufactured in a palaeo-technic age.

The George Edward Bellis vertical compound high-speed engine of the 1890s, for example, designed in Birmingham, required a newly patented revolutionary approach to lubrication: oil was forced under pressure direct to the surfaces requiring it. Lubricating oils derived from petroleum had begun to be used thirty years before, and by the end of the century it was clear that the use of modern high-speed machinery depended as much on lubrication as on precise measurement.

Another remarkable engine, P. W. Willans's central valve engine, constructed in the same year as Parsons's turbine patent, was highly original and equally ingenious. It employed tandem compound vertical cylinders, single-acting to avoid reversals of stress, combined with a unique form of steam distribution which saved space. Many were installed in power stations in the London area.

This period represented the swan-song of the reciprocating engine for electricity power generation, because reciprocating steam engines could not be made to run faster than about 500 rpm even when all enclosed and lubricated as the Bellis and Willans engines were, and larger engines had severe vibration problems. Nevertheless, reciprocating engines continued to be built and used for other purposes not requiring high-speed running until well into the 20th century.

Beneath the raised cover of a marine steam turbine, an engineer starts on the task of cleaning the 257 blades in one of the four turbine sets. The fixed guide blades through which steam is introduced into the turbine can be seen in the underside of the cover.

Steam for Pianoforte

Such was the novelty of the steam engine that it developed its own sub-culture of popular comic songs. The covers of a few of them are illustrated on this page.

One of the most fanciful songs was *The Steam Arm*, about a soldier who had lost an arm at Waterloo. 'The story goes on, that ev'ry night; His wife would bang him, left and right; So he determined, out of spite, To have an arm, cost what it might.' He had one made 'to work by steam'. When he got home 'and knock'd at the door, His wife her abuse began to pour;

He turn'd a small peg, and before He'd time to think, she fell on the floor. . . . For the soldier's arm had been so drill'd, that once in action it couldn't be still'd.'

More than one song explored the idea 'That a Rail-road to Mars, And the rest of the Stars, Will convey us direct to the Moon!' Indeed, steam would work all manner of marvels. 'Now Married Folk will have no trouble . . . For Children will by this wond'rous scheme, Sirs, Be Born and fed by the power of STEAM, Sirs. A STEAM education they will re-

ceive, too . . . and wou'd you believe . . . Make love by STEAM, and by STEAM get married.'

Love, if not marriage, also came into a song called *Bradshaw's Guide*. A man on a train meets a lady who 'could not recollect the town to which she wish'd to ride'. He helps her look for it in his *Bradshaw's Guide*, but she still has not found it when he has to alight. So: 'I took her to my quarters, where she took up her abode; And all the livelong day, then both of us, we tried to find the town she wanted in my *Bradshaw's Guide*.'

Two quadrilles, a march, a galop and a 'musical illustration' take a common theme, with scores by French, American and British composers.

QUADRILLE A VAPEUR

8

QUADRILLES JUSQU'A PARIS

COMPOSÉ PAR G. BAILLIEU

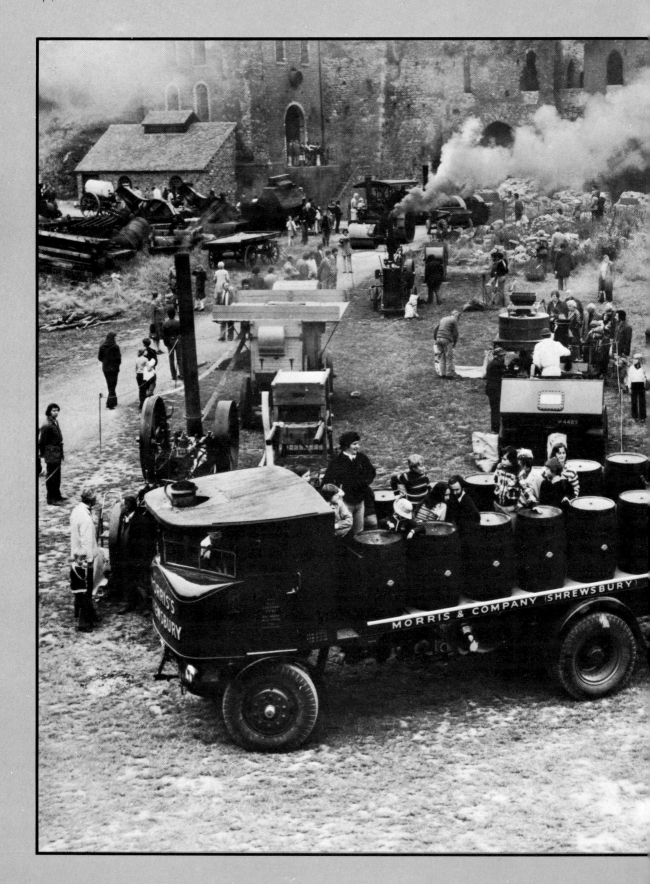

CHAPTER SIX

VINTAGE TECHNOLOGY

A period may arrive when even the steam-engine may be derided as an imperfect piece of mechanism, and some discovery made that will enable man to wield equal force without the employment of its cumbrous bulk and expensive fabrication.

ANONYMOUS, *A History of Wonderful Inventions* (1865)

There hasn't been a year like 1980 since steam preservation began. . . . There has been a succession of events that only a few years ago would have been beyond belief . . . from the humble volunteer has blossomed a full-scale steam Renaissance.

EDITORIAL, *Steam Railway*, No. 10 (Jan/Feb 1981)

At Blists Hill, Shropshire, visitors to a steam rally in 1977 examine a Sentinel lorry and a variety of traction and portable engines driving agricultural machinery. Blists Hill, until 1912 an iron-smelting complex, is now an open-air museum run by the Ironbridge Gorge Museum Trust, whose pioneering work in industrial archaeology has preserved the historic site at Coalbrookdale where Abraham Darby first smelted iron with coke in 1709 and, a little downriver, the world's first iron bridge, built in 1779.

The place of steam power in an age of alternative technologies was bound to be different from that when 'steam was king' and reigned supreme. The first real challenge to the supremacy came from electricity during the last two decades of the 19th century, although even earlier developments in gas technology, in the use of hydraulic power and in the exploitation of oil demonstrated that the quest for alternative forms of power was continuous.

Using the imagery not of monarchy but of slavery, a German writer, Arthur Wilke, in 1893 set the scene dramatically. 'Hardly had the slave *Steam* grown to its full strength, than there appeared for the service of mankind a young giantess, *Electricity*, who as it seems desires to work in harmony with Brother Steam for their masters, but in fact is proceeding completely to displace him. . . . No longer "the century of steam", no, the present period will be named "the age of electricity" '.

Through the invention of the turbine, steam and electricity were to work in harmony for their masters in the great electrical power stations of the 20th century. So, too, were water power and electricity, especially in Switzerland, where Geneva, enriched by generous water power, became the first big industrial user of electric power during the 1880s. Britain was slower to adopt electrical power than either Germany or the United States, where economic growth was faster. 'Clinging to steam', like clinging to old machinery, seemed to provide evidence of industrial decline: Arthur Shadwell in his *Industrial Efficiency* commented that Britain showed 'traces of American enterprise and German order, but the enterprise is faded and the order muddled'. There was no shortage of British inventions – even in electricity – but, as Charles Wilson has written, 'what was really serious' was that 'fewer inventions were being incorporated into industrial practice'.

The history of the economic uses of electricity began not with power, but with communications; the telegraph began to transmit Morse messages at the time when the greatest triumphs of steam, the railway and the steamship, were also transforming communications. Lighting by electricity came next (significantly, all the earliest British Acts of Parliament dealing with electricity were described as Lighting Acts). Driving machinery came last.

Europe's first public power station was opened at Godalming in Surrey in 1881. A *Punch* cartoon of that year showed King Steam and King Coal anxiously watching the infant Electricity and asking 'What will he grow to?' and the President of the Institute of Mechanical Engineers (founded in 1847, with George Stephenson as its first President) told apprehensive listeners that it was 'possible and even probable that one of the great uses to which Electric Force will be applied eventually will be the simple conveyance of power by means of large wires'. Seven years later, J. Munro, in the second edition of his *Electricity and its Uses* (published somewhat surprisingly by the Religious Tract Society), was still uncertain about the timing of future change, though he had no doubt that it would come. 'The steam engine, the telegraph, and coal-gas were the industrial triumphs of the past generation. It will be curious if electricity should not only serve the telegraph, but supplement the steam-engine and gas-lighting in the next. Its great advantage over all other sources of power is that it can travel to long distances in a moment, and thus act as a carrier of force as well as a stationary worker.'

During the 1880s a gospel of electricity began to be preached just as fervently as the gospel of steam had been during the previous decades. Electricity was said to have four advantages over steam: it had a wider range of uses (including not only lighting but industrial chemistry and medicine); it was more efficient in thermodynamic terms; it was cleaner; and it could transform both factory and home. It was associated from the start with science. Yet it had an aura of magic about it also. As the American Park Benjamin put it in his book *The Age of Electricity*, published in New York in 1887: 'Electricity is a vigilant and sleepless sentinel. . . . It will impel the locomotive; and equally, it will control the brake which stops its motion. . . . It will set, type and drive the printing press . . . and move the loom. It is already in use to control the warmth of the hatching egg: it has been proposed to use the current to cremate the bodies of the dead. . . . Where in the history of the magic are there wonders greater than these?'

Some writers brought out the superiority of electricity by emphasising the waste which seemed to go with steam – the novelist H. G. Wells was prominent among them – or by remarking, as Ruskin and William Morris had done, on the dirt and the smoke. Others were more positive, and looked forward to a time when domestic electrical devices in the home would take away drudgery as steam was said to have taken it away in the factory. (The main role of steam power in the home was to be the steam pressure cooker.) As for the factory powered by electricity, it would not only be cleaner than the factory powered by steam, enthusiastic social economists argued, but more easily managed. It need not be large, for the wide diffusion of electrical power would permit a large number of small factories rather than a few huge undertakings.

There was an element of exaggeration in such statements. Yet important differences did mark off 'carboniferous capitalism', as Lewis Mumford called it, from economies based on the use of electric power. Not only did 'the pall of palaeo-technic industry begin to lift', and clear skies and clear waters come back again; the development of electricity, leading in the late-20th century to the development of revolutionary electronics, rested on the final breaking down of what the 19th-century pioneering electrical industrialist William Siemens called in 1876 'the barrier between pure and applied science'. Moreover, electrification required planning and control by public utilities in a way that the development of steam power had never done. In some countries, notably the United States, such utilities were privately owned. In most others, however, they were in municipal or State hands. It is not surprising, therefore, that the gospel of electrification was taken over by Lenin in the Soviet Union after the Bolshevik revolution of 1917, or that electric power stations now appear (like agricultural tractors) on bank notes in communist countries.

Two of the most articulate enthusiasts of electrical power lived in the United States: well-born Henry Adams (historian, philosopher and grandson of a President), who compared the cult of the dynamo in the 19th century with the cult of the Virgin in the 13th; and Nikola Tesla, inventor and prophet, who was born in Croatia of Serbian parents.

Adams contemplated with awe the dynamos which were on display at the Great Chicago Exhibition of 1893 and the Paris Exhibition of 1900 (where there was a whole Palace of Electricity). Professing 'the

An electric roller built in the early 1900s by Thomas Green & Co. Leeds (and probably driven by Thomas Green himself) looks not much different from the heavy iron horses of the steam age. It was powered from an overhead pick-up. Though not produced in quantity, it exemplified the challenge to steam posed by alternative technologies.

In the twilight era of the early 1900s, before
motor vehicles had become well established, a
steel-tyred Robey steam wagon is put through its
paces. A vertical boiler supplied steam at 200
pounds per square inch to a compound engine
mounted underneath – a so-called 'undertype
engine'. Steam, with its great tractive power,
was better at hauling heavy loads than providing
convenient, fast passenger transport, and steam
lorries were in use as late as the 1940s.

religion of World's Fairs', he compared the power of electricity with the
power of the Cross. Describing the 20th-century American as 'the child
of steam and the brother of the dynamo', he even brought the atom
into his range of vision:

> 'Seize, then, the Atom! rack his joints!
> Tear out of him his secret spring!'

Tesla, who worked in Paris before moving to the United States in
1884 with four cents in his pocket, had more wild fantasies about
electricity than Trevithick had about steam.

Between the first and last years of the 20th century, while many more
countries have become industrialized, three main changes have taken
place in the history of power, some of them packed into the two most
recent decades. These changes place steam power in yet another
perspective. They have been so sweeping that some interpreters, like
Peter Drucker and Alvin Toffler, have described them not only as a
historical discontinuity but as a social and cultural mutation. In
particular – and this is the first big change – the post-Second World War
development of electronics (with Japan to the fore) has altered the
balance of industry. The development has gone through different
phases, bringing with it new techniques, the sense of a new range of
human problems and possibilities, and even, it has been claimed, a new
consciousness. 'If a technology is introduced from within or without a
culture,' Marshall McLuhan has written, 'and if it gives new stress or
ascendancy to one or another of our senses, the ratio between all the
senses is altered.'

The now huge electronics industry depends on scientific research and
development. Computers have already passed through several 'generations' in little more than thirty years, the space of a human generation,

Though outwardly much like any petrol-engined car of its period (right), this American-built Stanley steam car of 1920 used its radiator as a condenser, and under the bonnet (below) it had a flash boiler heated by kerosene. The white part is asbestos lagging to speed up heating, which took six or seven minutes from cold. The car had a top speed of 55–60 m.p.h., and similar vehicles were produced by the Stanley company until 1927.

and have made it possible to achieve the kind of results, particularly automation, which Andrew Ure dreamed of during the early history of steam power. Even Ure, however, did not forecast the conquest of space, which would have been impossible without electronics. Jules Verne, who did, was so much caught up in his own age that his imaginative invention still focussed on steam (as in his novel *The Steam House*, published in 1880) and on electricity. Since the Second World War, there have been as many prophets of the new computer age as there were disciples of the gospel of steam. There is a difference, however, in orientation. Until it was challenged in the 1960s, the 20th-century conception of the role of science rested not on diffusion and popularization – making knowledge accessible to all, which was the main feature of the age of steam – but on the education and formation of a well-trained scientific elite. *The English Mechanic*, which first appeared in 1865, had talked of spreading knowledge of steam power among 'the millions'. Knowledge of electricity – and later of nuclear power – could

From the graveyards of steam, such as that (far right) at Edwards' Yard, Swindon, in 1957, preservationists salvaged what they could. Among them was the enterprising Beamish North of England Open Air Museum, which in 1968 took delivery of a colliery ventilating fan (right).

The last days of steam were tinged with the sadness of fine design and craftsmanship turning to dereliction. A good example of this plight is the magnificent interior of Crossness Sewage Pumping Station (below), at Erith on the south-east edge of London, where dust and cobwebs adorn the cylinders and iron tracery of two of the four rotative beam engines. Originally single-cylinder engines, installed by James Watt and Sons in 1864, they were converted to triple-expansion working in 1901, and stopped in 1945.

not be spread in this way. Another difference is that the new 20th-century technologies displace labour on a far bigger scale than steam technology ever did.

The second great change is the development of the oil industry and of the internal combustion engine, a story that began in the last decades of the 19th century. 'Your machine inflicts a blow on the powerful steam engine,' Franz Reuleaux wrote to Rudolf Diesel in 1983, 'since it exceeds it in heat efficiency. Technology must ultimately eliminate the fault so long recognised in the old steam engine. The increase of power that we can gain thereby is so great that it is worth every effort.'

The power shift, which had begun with automobiles, continued with aeroplanes. Although Erasmus Darwin's dream of the conquest of the air by steam was never realized, understanding of some of the principles behind powered flight had reached a sophisticated level when the age of steam was still young. Sir George Cayley, founder in 1831 of the British Association for the Promotion of Science, had already identified the aerodynamic forces operating on a wing in 1799 before Trevithick played Catch-Me-Who-Can. 'He knew more of the principles of aeronautics,' wrote Orville Wright, pioneer of the 20th-century aeroplane, in 1912, 'than any of his predecessors, and as much as any that followed him up to the close of the 19th century.' Interested in bicycles as well as aeroplanes, Cayley recognised that the sky offered new possibilities: 'an uninterrupted navigable ocean that comes to the threshold of every man's door, ought not to be neglected as a source of human gratification and advantage'.

Britain lagged behind other countries in the first chapters of the story, as internal combustion engines competed against horses – the first petrol-driven car in the country was imported from Germany in 1894 – just as it did in the later chapters when the motor car was transformed from a luxury good to a family 'necessity'. Nevertheless, motor cars were greeted in Britain with rhetoric no less enthusiastic than that which had accompanied the arrival of steam railways. At a motor exhibition in 1895, Harry Lawson, founder of the Great Horseless Carriage Company, told his audience that they were 'celebrating the birthday of the most wonderful industries that God has ever blessed mankind with since the world began'.

From the start, motor cars had their critics, too, some objecting to the image of luxury with which they were associated until Henry Ford brought in mass production (whereas early steam railways had been hailed as 'democratic'), others objecting to the way in which they disturbed the peaceful countryside as they raced with 'incredible velocity and no apparent aim' down winding country lanes. Aeroplanes also were to have a controversial history, unlike their predecessors, balloons. The element of danger and of harm was always present. Even more important in determining attitudes, air power became a major factor in war, drawing in civilians as well as soldiers. The latest technical triumph, the supersonic Concorde, has always been controversial socially and culturally as well as economically – at least as much as the steam engine ever was. Its development was more costly – and the costs more difficult to forecast – than any other aeroplane, and on both sides of the Atlantic it divided people who admired its grace from people who feared its environmental consequences.

The development of the internal combustion engine was so rapid that its long-term effects on railways and on steam transport by sea were not appreciated at once. The great growth of civil aviation, still not fully charted by historians, did not come until after the Second World War, allowing the steamship its elegant inter-war heyday. The challenge to railways and steam power on land had by then already been posed. In Britain between 1919 and 1939 the number of motor vehicles on the roads had risen nearly tenfold (cars increasing twentyfold and goods vehicles fivefold). These figures were dwarfed in the United States. In 1939, when just over two million motor vehicles were registered in Britain, the United States had more than 26 million. Thousands of internal combustion engines began to be employed on the land also, and the tractor became as much a symbol of progress as the steam plough once had been. (It, too, has figured on the banknotes of communist countries.) The switch from steam to oil, which began when the first Diesel locomotive was tried out in the United States in 1924, did not save the railways. It was no consolation to railway champions that during the 1970s there was talk of an automobile and oil crisis in addition to the railways' problems.

This crisis is often thought of (rightly) as a global crisis which fundamentally changes – or will change – both economic relationships and ways of life. Decades ago the great French historian Marc Bloch wrote that 'the man of the age of electricity and the aeroplane feels himself far removed from his ancestors'. Yet Bloch, who was killed by the Nazis, did not live long enough to see the third great change of the 20th century: the development of nuclear power, which followed from the expensive Allied atomic bomb programme during the Second World War – the first power technology to come out of a research programme based entirely on war needs. It was clear long before the completion of the world's first nuclear power station at Calder Hall in Britain in 1956, or the growth of American atomic power plant in business hands, that

At an Oxfordshire farmer's open day in 1979, a before-and-after display includes (second in line) a finely restored Marshall engine made and exhibited at the Smithfield Agricultural Show in 1887. The preservation movement began after 1952, when another local farmer, Arthur Napper, challenged the Marquis of Bath to a traction engine race, arousing interest that saved many machines from the scrap heap.

A dinosaur from the age of steam, a mahogany-burning Shay locomotive built in Ohio in the early 1900s showers sparks as it grinds over rickety rails with a trainload of timber in the Philippines. Patched together and kept going by a lumber company, it was still working in 1975, when many of the monuments of steam wore the aspect of an engine house (above) near Pontypool in South Wales. This rotative beam engine was built by the Neath Abbey Engineering Works in 1845, and has been derelict for half a century.

A 1913 Fowler engine, dredging near Sheffield in 1968, blows off steam before winching in the mud scoop. Dredging was one of steam's last active tasks.

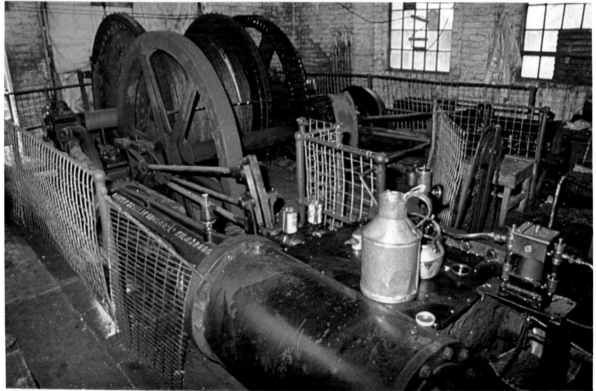

A colliery haulage engine, driving a continuous wire rope to which coal tubs were attached underground, still has the splendour of working steam in 1976.

there would be peaceful as well as warlike uses of nuclear energy. Yet by the 1970s, the peacetime development of atomic energy had become no less controversial in countries as different as Sweden and France, the United States and Japan than the atomic bomb – and later the hydrogen bomb – itself had been.

Given the three main technological trends of the earlier 20th century, steam power still plays a much more active part than it might have been expected to play. First, it continues to be used in old ways in a few parts of the world, including Zimbabwe, where Beyer-Garratt locomotives are being kept in operation; China, which is still building steam locomotives; India; the Philippines, where steam engines are at work in the sugar fields; and South Africa, where the phasing out of steam locomotives, decided on in 1971, will not be complete until the 1990s. Second, following the global energy crisis, there has been a renewed interest in the possibilities of coal rather than of oil and of steam power (along with the older technologies of water and wind power) rather than of controversial nuclear power, and in some cases decisions to 'scrap' steam have been revised. Third, 'vintage' steam has attracted an increasing number of enthusiasts in many countries: indeed there are probably more steam enthusiasts scattered round the world now than there were during the age of steam itself, when the full gospel was being proclaimed.

Each of these features is regularly discussed in newspapers and periodicals, just as the potential of steam was discussed during the 18th and 19th centuries: there are also specialist magazines on steam power, many of them superbly illustrated, and books that capture the last days of steam wherever it is still to be found. The sense that the last steam engines are in danger provokes something of the same feeling as threats to wild life can do. Indeed, the German Tourist Office in London publishes periodical newsheets, like lists of endangered species, on 'pockets of steam' that still survive in Germany.

The second feature – involving the exploration of 'energy alternatives' – has been discussed time and time again by national and international agencies. It forms part of a global pattern. In 1975 industrial countries with only twenty per cent of the world's population consumed five times more total energy and twelve times more energy *per capita* than the developing countries. Such enormous disparities, which had their origins in the age of steam, are increasingly coming under attack. If industrial countries are to reduce their reliance on oil, they may find it feasible to make selective use of steam fuelled by coal in the future.

Steam may still be used in partnership with other forms of power – as steam turbines have been employed in conjunction with water turbines to even out peaks and troughs in electricity loads – and new uses of steam are often being considered. China, indeed, has a steam engine research programme. Although more attention is usually paid by energy specialists to the potentialities of hydro-electricity, wind and water power, solar energy and geothermal power, not to speak of biomass, there are vocal advocates of 'steam at the turn of a switch' for road and railway transport. A *Light Steam Power Magazine*, for example, advertises a list of dimensional drawings for a monotube steam generator, a steam marine plant and a light steam car power unit among other things, while

One of the many preserved steam railways in the United States advertises a five-mile round trip to a disused silver mine near Denver, Colorado.

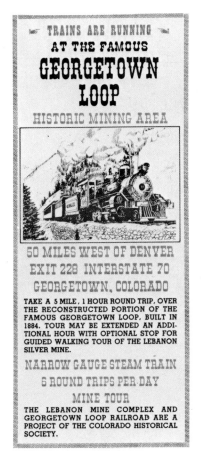

TRAINS ARE RUNNING
AT THE FAMOUS
GEORGETOWN
LOOP
HISTORIC MINING AREA

50 MILES WEST OF DENVER
EXIT 228 INTERSTATE 70
GEORGETOWN, COLORADO

TAKE A 5 MILE, 1 HOUR ROUND TRIP, OVER THE RECONSTRUCTED PORTION OF THE FAMOUS GEORGETOWN LOOP, BUILT IN 1884. TOUR MAY BE EXTENDED AN ADDITIONAL HOUR WITH OPTIONAL STOP FOR GUIDED WALKING TOUR OF THE LEBANON SILVER MINE.

NARROW GAUGE STEAM TRAIN
6 ROUND TRIPS PER DAY
MINE TOUR

THE LEBANON MINE COMPLEX AND GEORGETOWN LOOP RAILROAD ARE A PROJECT OF THE COLORADO HISTORICAL SOCIETY.

Dr. John Sharpe of Queen Mary College, London, has advocated for certain purposes new locomotives, which would look rather like diesel locomotives and, instead of puffing steam into the air, use a turbocharger which converts it back into water and back again into steam: they would, he argues, be pollution-free and would use logs and wood chippings instead of coal just as American paddle-steamers did in the 19th century. In Argentina also a railway engineer has put forward practical proposals for a modernised form of steam locomotive.

The appeal of vintage steam is not just a symptomatic expression of British nostalgia for times past. Although the age of steam was Britain's hour of industrial supremacy, there can be as much enthusiasm for vintage steam in Europe, the United States and even in Japan, pioneering centre of the new technologies. In the United States, where every new technology, however fanciful, gets a hearing, the paddle-steamer *Mississippi Queen*, combining 'modern standards of comfort and convenience' and traditional steam engines with 'ante-bellum splendour', still makes its way from New Orleans to Natchez with the calliope playing tunes like *Steamboat Round the Bend*. (An article on a *Mississippi Queen* cruise by Wim Van Leer recently appeared as far away and in as unlikely a place as *The Jerusalem Post*.) One of the few jarring notes in the new symphony of steam came from Britain: the *Oxford Mail* in October 1980 described Didcot families protesting to the local council about the noise (particularly the whistles) of steam engines at Didcot Steam Centre: the heading, worthy of Nathaniel Hawthorne, read 'Steamed up over rail fans' whistle?'

'Farewells to Steam' have, of course, been many. Farms seldom ritualized them. Nor did fairgrounds, where diesel prime movers came to replace steam engines unobtrusively. But on the railways, as with trams, the farewells were often fulsome. As told by stirring writers like O. S. Nock, the 20th-century story of railways was essentially romantic, often focussing on the Indian summer of steam during the 1950s. Thus, when in Australia the Victorian Railways North Loco Steam Depot, built in 1888, was closed in 1965, it is said to have died in a blaze of glory. 'The working lives of a thousand railwaymen centred around it,' P. A. Smith has written. 'When the time came for it to be demolished, all the old drivers and firemen turned up to watch it fall. It didn't go out with a whimper.'

'Have you noticed how nostalgia grips us all at times?' the editor of the British magazine *Railway Reflections* asked in 1980. 'Whatever the subject, whether we're young or old, we tend to reflect back on what has been and never will be again.' This desire was particularly strong in 'the world of transport', he claimed, 'because transport is a visible world, and yet is rapidly changing – with constant improvement and supercession.' 'Railways,' wrote the novelist Paul Theroux more simply, 'are irresistible bazaars.' Accounts of his own journeys through little-explored parts of the world have been bestsellers.

Nostalgia can soon dissipate: preservation involves positive and continuous action. The railway steam preservation movement in Britain is said to have begun in the 1950s – after nationalisation – with the acquisition by private groups of the moribund Talyllyn and derelict Ffestiniog railways in Wales. The first, in 1950, carried no purchase price and the latter, in 1955, a small one. Nor were the first locomotives

In the cab of a Dean goods locomotive at the Great Western Railway Museum, Swindon, a driver and stoker's eye view captures all the fascination of preserved steam. In the centre, between the steam pressure and vacuum gauges, is the steam equivalent of an accelerator, the regulator. When opened slightly, it operates a jockey valve to keep steam on the lubricators when the engine is coasting. To the left of the regulator is the water gauge, and to the right the vacuum brake, with its air filter by the right spectacle plate. Hanging above is the chain for one of the two whistles. The reversing lever is on the far right, next to one pair of suspension springs for the trailing wheels. The crew had to be careful not to get their trousers caught in the springs when the engine was bouncing over a rough stretch of track.

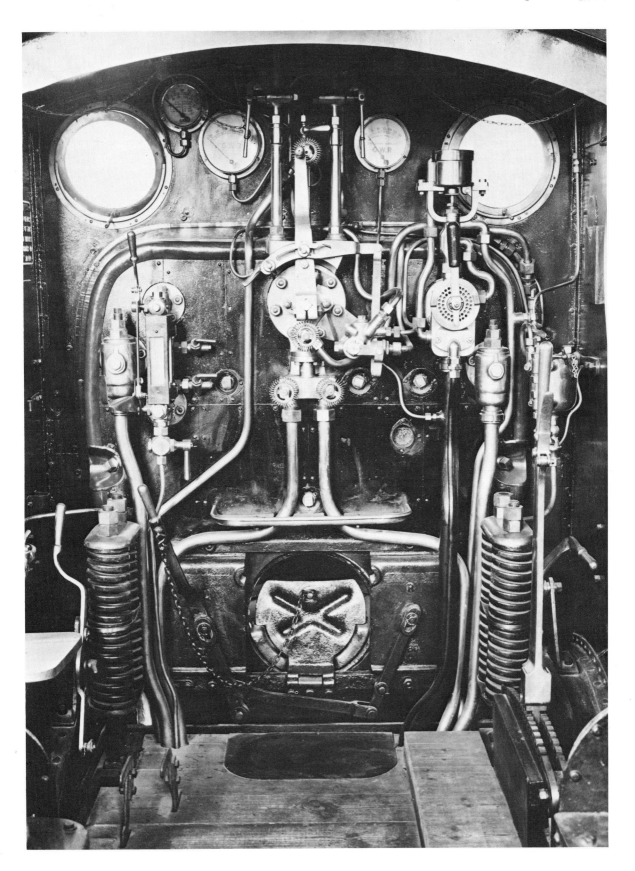

expensive: the Talyllyn's *Sir Haydn* (built in 1878 and named after the former owner, Sir Henry Haydn Jones) was bought from British Railways Western Region in 1951 for only £25. The Bluebell Railway in Sussex was founded in 1960, and in the same year the Middleton Railway near Leeds, which until then had carried coal, was also acquired by preservationists.

It was not only railway track and locomotives which interested preservationists: they were said to be interested in 'steam in aspic' of all kinds. Austria had a national transport museum in the late-19th century and was already requiring its railway companies to send in historical items for preservation; and in the 20th century Germany created a superb railway museum near the station in Nuremberg. Railway relics were being collected in Britain as long ago as 1925,

A Rescue Operation

A Ruston-Proctor steam navvy worked at Arlesey chalkpit, Bedfordshire, from 1909 to 1931, when the pit was abandoned. Water gradually filled the pit, submerging the navvy, but in 1976 a steam rescue enthusiast, Ray Hooley, noticed the jib poking up. A year later the navvy was raised. It was restored by apprentices and steamed again in 1980, with George Albon, the last driver in 1931, at the controls.

A Ruston-Proctor steam navvy cuts chalk at Arlesey, Bedfordshire, in 1921. Among the workers (inset) is George Albon, seated front row, second from left.

when the centenary of the Stockton and Darlington Railway was celebrated, and at the 150th anniversary celebrations of the same line in 1975 the National Railway Museum was opened in York.

Exhibits from the Museum of British Transport, opened in 1963 near the great Clapham railway junction in South London, were moved north (not without controversy), and the first Keeper, associated with the Science Museum in London, described his object as erecting a 'living museum', exhibiting both steam locomotives and all forms of motive power. He also collected objects, paintings, posters and photographs, and by the belated Transport Act of 1968 acquired first claim on nominated items from British Railways once they ceased to be in regular use. The locomotives now housed at York include the last steam locomotive to be built for British Railways in 1960, appropriately named

The navvy's jib rises from the flooded chalkpit.

Mud encrusts the winch of the retrieved navvy.

The navvy steams up for the first time in 49 years. A new corrugated iron roof stands ready alongside.

After restoration, the main engines and controls are tested on compressed air in the training department of Ruston-Bucyrus, successors to Ruston-Proctor.

Evening Star.

Official action is supplemented in every country by the action of volunteers. Thus, in Britain, where there is a thriving Association of Railway Preservation Societies (along with a Transport Trust), the steam shed at Didcot, criticised in the *Oxford Mail* for its whistles, had become the centre of the Great Western Society in 1967, the year that David and Charles published the first volume of the invaluable work of reference *The Railway Enthusiasts' Handbook*. The top ten steam lines include the Severn Valley Railway based at Bridgnorth, Shropshire; the Bluebell Railway in Sussex; the Kent and East Sussex Railway at Rolvenden; The North Yorkshire Moors Railway near Pickering; and the Spey Valley Line in the Scottish Highlands. Depots are repaired too. In 1969 an old locomotive depot at Tyseley near Birmingham became a railway museum. Britain's last active steam shed, which continued in regular use until 1968 at the railway junction of Carnforth, was later taken over by a private company and given the proud, symbolic name of 'Steamtown'. Dickens's Coketown had been left far behind: Coketown suggested horror and routine, Steamtown delight and adventure. The *Flying Scotsman*, bought by a private individual in 1963, is one of the famous locomotives housed there.

Railway enthusiasts in Britain flock to York, in particular, as a place of pilgrimage. Arriving on British Railways' inter-city 125s they can see steam in action there on the main lines as well as preserved steam in the museum. The age of regular steam ended at York in 1967, when diesel 'Deltics' replaced the old LNER steam 4–6–2s, but there are now frequent 'steam specials'. The great station itself provides a perfect 19th-century setting.

My own birthplace, the town of Keighley, not far from York, is now a steam centre too. It is as much a product of the industrial revolution as York was of the Roman occupation, and its mills once had as many steam engines in operation as York had locomotives. The railways, however, are what have turned it into a centre. The motto of the town is 'By Worth', and it was on the small four-and-a-half-mile, run-down Worth Valley Line, closed in 1962, that volunteers began to move steam trains again six years later. Among the six stations on the old local line is the Haworth of the Brontës.

Few modern readers know that Charlotte Brontë was interested not only in railway steam, but in the power of steam applied far more generally. One contemporary critic of her novel *Shirley*, published in 1849, condemned her for using Northern colloquialisms like 'Miss Mary, getting up steam in her turn'. Yet Charlotte could rise to poetic heights with steam. When the hero of *Shirley*, Robert Moore, whose mill is his Moloch, is asked by the most Yorkshire of Yorkshiremen, Mr. Yorke, 'What has gone wrong?', he replies: 'The machinery of all my nature, the whole enginery of this human mill: the boiler, which I take to be heart, is fit to burst.'

'That suld be putten i' print,' Mr. Yorke responds with a touch of irony, 'it's striking. It's almost blank verse. Ye'll be jingling into poetry just e' now.'

It is the poetry of steam which persists: it has indeed often been more forceful in the 20th century than in the 19th. The Italian 'Futurist' poets, whose leader, Marinetti, accused Ruskin of 'infantilism', were equally

A close-up of a 6000's front end reveals details of the unique leading truck of the Great Western Railway King Class, combining inside bearings to the rear axle and outside bearings to the leading axle. Also evident is the horizontal rocking shaft actuating the valves of the outside cylinder. The inner end of the rocking shaft transmits motion from the Walschaert's valve gear of the inside cylinders, located beneath the smokebox.

impressed by automobiles and aeroplanes on the one side and on the other by 'adventurous steamers that sniff the horizon' and 'deep-chested locomotives whose wheels pound the tracks like the hooves of enormous steel horses bridled by tubing'.

Decades later in Britain, Cecil Day Lewis was equally attracted (as were leading documentary film makers) both by pylons and steam locomotives:

> 'Let us be off! Our steam
> Is deafening the dome . . .
> Valves cannot rent the strain
> Nor iron ribs refrain
> That furnace in the heart.
> Come on, make haste and start
> Coupling-rod and wheel,
> Welded of patient steel,
> Piston that will not stir
> Beyond the cylinder
> To take in its stride
> A teaming countryside.'

Stephen Spender published a poem 'The Express' in 1933, and seven years later W. R. Rodgers's poem with the same title forged his metaphors about the body in terms of steam:

> 'the through-train of words with white-hot whistle
> Shrills past the heart's mean halts, the mind's full stops
> With all the signals down.'

These images belong to the last age of steam on the railway. Subsequently, the poetry of motion itself – and of the power of steam – has moved industrial archaeologists as much as railway enthusiasts, directing them to 'industrial remains' of every kind from the age of steam. It is very much a poetry of the past, but it is a poetry which is concerned with very specific people, things and events.

With the poetry of steam there can go music – and art. Records of the sound of 'by-gone' steam include the sound of steam at the fair as well as on the railway – with one record called *Steam in the Worth Valley*, another *The Power of Steam* and a third *Requiem for Steam* 'by a onetime footplateman'. One branch of the art of steam is covered in books like *The Art of Railway Photography, A Pictorial Manual of Photo-Technique*, another in the continued production of anniversary plates, mugs and coasters dealing with steam themes, and railway company ties. Brass models of the *Rocket* are on sale as is a *Rocket* goblet -- and there is a flourishing model railway hobby group.

The poetry, the music and the art, however, are not perhaps at the core of the interest in vintage steam. Making steam work, as it always did, requires workers; and among most contemporary steam enthusiasts there is a delight in work itself, all the more conspicuous in that ours is an age when the gospel of work is unfashionable. Whereas collectors seldom leave their homes, except on the quest, steam enthusiasts revel not only in meetings and rallies but in actually working on the machinery – and it is dirty work. Peter Kelf, the editor of *Steam Railway*, summed up the still persuasive power of steam when in January 1981 he saluted the volunteers who 'scrape muck from rusting hulks and turn out in all weathers to ensure that steam will rise again'.

In a yard behind oast houses in Kent, a contractor's engines get up steam for the day's work. Various makes are represented, including (right) a large ploughing engine with its winch visible beneath the boiler.

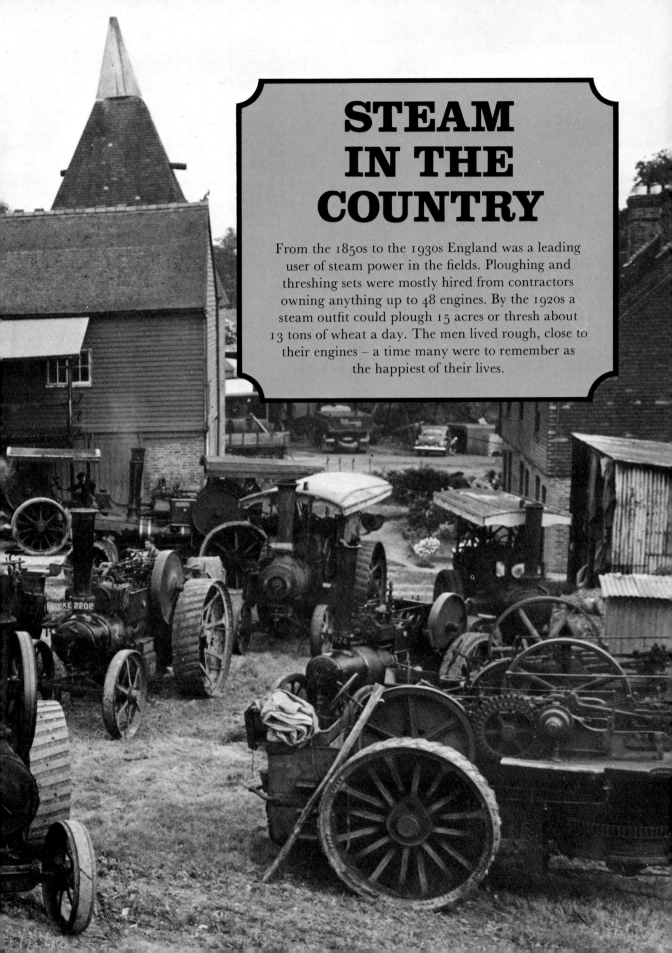

STEAM IN THE COUNTRY

From the 1850s to the 1930s England was a leading user of steam power in the fields. Ploughing and threshing sets were mostly hired from contractors owning anything up to 48 engines. By the 1920s a steam outfit could plough 15 acres or thresh about 13 tons of wheat a day. The men lived rough, close to their engines – a time many were to remember as the happiest of their lives.

Guided by a steersman, a balanced plough is winched through a heavy clay by a distant engine, one of a Fowler two-engine set straddling the field.

Tilling the Soil

Powerful traction engines weighing up to 18 tons could not plough heavy soils directly, but a system of cable ploughing (left), developed by Fowlers of Leeds, was very successful. In 1900 about 600 cable sets were in use in Britain, and many abroad. But they required five-man teams and were costly to buy. By the 1930s, like most other steam cultivating machinery, they had mostly had their day.

Tomato growers fill a soil sterilizer in Guernsey.

A two-man crew plough a field with a Suffolk Punch light steam tractor, introduced in 1917 by Richard Garrett & Co. to meet the threat of the motor tractor.

Sawn timber piles up in an English wood as logs are fed to a circular saw driven by a portable engine. The power is free, since the engine burns offcuts.

The Colonist, a vertical-boiler road steamer of a type built for India by Ransomes of Ipswich, transports a tree for one of the firm's partners in 1872.

Farmer's Choices

Steam engines did many jobs in the country apart from cultivating and threshing. They drove saw mills and water pumps, sterilized contaminated soil (a process still used in Guernsey), and dredged, ditched and drained the land. On country estates small stationary engines often powered flour mills, dairies and workshops. Traction engines could haul everything from ploughmen's living cabins and heavy timber to wagonloads of children on outings.

An apple-picking party in a Foden steam wagon sets out to a Buckinghamshire farmer's orchard in 1917.

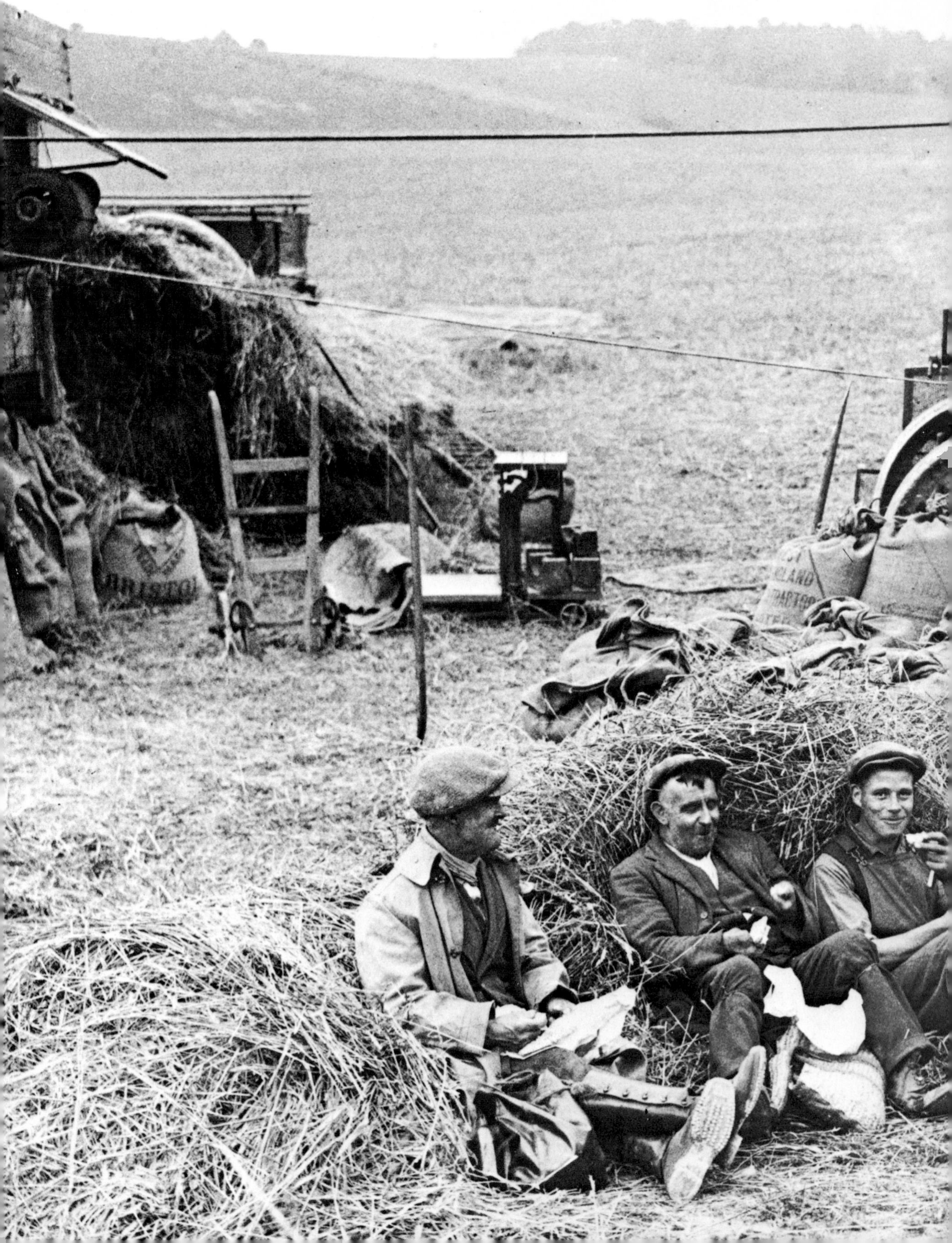

Threshers stop for lunch at a farm in Oxford-shire in the early 1920s. The sacks of corn they have filled with their Fowler engine and threshing machine make a satisfying backrest.

NOTES ON SOURCES

Many books have been consulted in the preparation of this volume. These notes do not cite them all. They are a selective, critical readers' guide to a huge literature. Particularly useful and readable books are starred.

INTRODUCTION

There are several histories of technology which pay varying degrees of attention to the economic, social and cultural significance of the steam engine. .

Among them are the Oxford *History of Technology*, especially Vol. IV, *The Industrial Revolution* (1958) and Vol. V, *The Late Nineteenth Century* (1958). A useful French survey, translated into English in 1979, is M. Daumas (Ed.), *A History of Technology and Invention*, while I have also found useful the German study, also translated into English, F. Klemm, *A History of Western Technology* (1959). See also L. T. C. Rolt, *Victorian Engineering** (1970).

For broader and more speculative studies, see A. P. Usher, *A History of Mechanical Inventions* (1934); L. Mumford, *Technics and Civilisation* (1934); L. White, *Medieval Technology and Social Change* (1962); J. U. Nef, *The Conquest of the Material World* (1964). N. Rosenberg, *Perspectives on Technology** (1976) is particularly stimulating.

For science and steam, see A. E. Musson and E. Robinson, *Science and Technology in the Industrial Revolution** (1969); D. S. L. Cardwell, *Steam Power in the Eighteenth Century* (1963); S. Carnot, *Réflexions sur la Paissance Motrice du Feu* (1824); S. Lilley, 'Attitudes to the Nature of Heat about the Beginning of the Nineteenth Century' in *Archives Internationales d'Histoire des Sciences*, Vol. 27 (1948); and Y. Elkana, *The Discovery of the Conservation of Energy* (1974).

For engineering, see A. F. Burstall, *A History of Mechanical Engineering* (1963) and W. H. G. Armytage, *A Social History of Engineering** (1961).

There is excellent illustrative material in many books, including F. D. Klingender, *Art and the Industrial Revolution** (revised edn., 1968) and more recently J. T. van Riemsdijk and K. Brown, *The Pictorial History of Steam Power** (1980).

German historians have proved themselves more aware of the cultural ramifications of the subject than the British or Americans. See *inter alia* C. Matschoss, *Geschichte der Dampfmaschine* (Berlin, 1901). See also L. Marx, *The Machine in the Garden* (1967) and H. L. Sussman, *Victorians and the Machine* (1968).

There have been many periodicals in different countries concerned with the uses of steam power throughout most of its phases – among them the American *Steam Engineering* (New York, 1886); the German *Dampf: Organ für die Interessen der Dampf Industrie* (1888); and *The International Steam Engineer*, founded in 1896, which produced an interesting account of fifty years of progress in 1946. *New Steam Age* was launched in Connecticut as late as 1942, and *Light Steam Power* first appeared in 1949.

Amongst a vast number of books on engines I found particularly useful R. J. Law, *The Steam Engine** (1965) and the American Anchor Book by J. F. Sandfort, *Heat Engines** (1962), which includes in its introduction the reassuring sentence 'thermodynamics is not the stodgy and abstruse subject that many suppose, but rather a fascinating science that grew out of a great human need – the production of power from heat'.

CHAPTER 1

The best introduction to the themes of this chapter is L. T. C. Rolt and J. S. Allen's *The Steam Engine of Thomas Newcomen** (1977); it has a useful bibliography. See also R. Jenkins (Ed.), *The Collected Papers of the Newcomen Society* (1936); and the immensely valuable articles scattered throughout the society's *Transactions*.

There are several 19th-century histories of the steam engine, which cover the period discussed in this, the next and some of the following chapters. They include: J. Farey, *A Treatise on the Steam Engine** (1827); T. Tredgold, *The Steam Engine* (1827); D. Lardner, *The Steam Engine Explained and Illustrated** (1828); R. Stuart (pseudonym), *A Descriptive History of the Steam Engine* (1824); E. Galloway, *The History and Progress of the Steam Engine* (1839); R. H. Thurston, *A History of the Growth of the Steam Engine** (1878); R. L. Galloway, *The Steam Engine and its Inventors** (1881); and Sir George Holmes, *The Steam Engine and other Heat Engines* (1895).

Twentieth-century accounts include H. W. Dickinson, *A Short History of the Steam Engine** (1938) and J. D. Storer, *A Simple History of the Steam Engine* (1969).

J. Needham, 'The Pre-natal History of the Steam Engine' (*Newcomen Society Transactions*, Vol: XXXV*, 1963) offers fascinating suggestions about Chinese development.

History Today, Vol. 30 (1980) includes four interesting articles on pre-steam sources of power, including 'Wind power' by W. Minchinton, to whom I am indebted for introducing me to the excellent unpublished Exeter University doctoral thesis by J. W. Kanefsky, *The Diffusion of Power Technology in British Industry* (1979). This is a subject dealt with also in G. N. von Tunzelmann, *Steam Power and British Industrialization to 1860* (1978).

See also W. Rees, *Industry before the Industrial Revolution* (1968) and R. Hills, *Power in the Industrial Revolution* (1970).

CHAPTER 2

H. W. Dickinson, *James Watt, Craftsman and Engineer** (1936) is the standard account. See also his and R. Jenkins's *James Watt and the Steam Engine** (1927). These volumes should be supplemented by E. Robinson and D. McKie, *Partners in Science, James Watt and Joseph Black* (1970) and E. Robinson and A. E. Musson, *James Watt and the Steam Engine** (1969). See also H. W. Dickinson, *Matthew Boulton* (1936) and T. H. Marshall, *James Watt* (1925). A useful technical study is R. J. Law, *James Watt and the Separate Condenser* (1976).

J. R. Harris wrote a pioneering article on 'The Employment of Steam Power in the Eighteenth Century' in *History*, Vol. LII (1967).

Watt's achievement rang round the world during the 19th century and continued to ring round it a century later. See, for example, *The James Watt Bicentenary Commemoration Volume, being a Record of the Celebrations held by New Zealand Engineering Societies* (1936), published by New Zealand electrical engineers.

CHAPTER 3

The gospel of steam was propounded in many lectures, sermons, books, pamphlets and articles which were not concerned exclusively with steam. A. Ure, *The Philosophy of Manufactures* (1835), is essential central reading, as is his powerful critic Karl Marx's *Capital* (1867). See also S. Smiles, *Lives of Boulton and Watt* (1865) and L. Figuier, *Les Merveilles de*

la Science (Paris, 1867). One of the most interesting autobiographies is that of *J. Nasmyth** (1891): it was edited by Smiles. Anon, *The Triumphs of Steam* (1858) was dedicated to Robert Stephenson. See also R. H. Thurston, *The Development of the Philosophy of the Steam Engine* (New York, 1888), and J. T. Ward, *The Factory System*, two vols. (1970).

The idea of *Steam for the Millions* was propounded explicitly in Philadelphia in 1847, although as early as 1805 Oliver Evans had produced in the same city *The Young Steam Engineer's Guide: An Investigation of the Principles, Construction and Powers of Steam Engines*. Such guides became and remained common in Britain's industrial provinces. See, for example, J. Hopkinson and others, *The Engineer's Practical Guide* (Huddersfield, 1875). They were common in other countries too. My favourite title is American – H. E. Collins's *Knocks and Kinks; Causes, Detection and Cure for Many of the Commonest Troubles of the Engineer* (New York, 1908).

Some authors wrote books at two levels for advanced students and beginners, the latter often including 'gospel' material. Thus, D. Lardner wrote a *Rudimentary Treatise on The Steam Engine for Beginners* (1848), while R. Stuart produced his *History and Descriptive Anecdotes of the Steam Engine* in 1829. J. Bourne, who produced a *Table of the Nominal Horse Power of Steam Engines* (1869), also edited for the Artisan Club *A Treatise on the Steam Engine in its Application to Mines, Mills, Steam Navigation and Railways* (3rd edn., 1850) and drafted his well-known *A Catechism of the Steam Engine* (1865).

Nothing suggests a gospel more than a catechism. including a French *Catéchisme des Chauffeurs et des Méchaniciens* by R. Marquis, which had gone through nine editions by 1924, and a German *Kateckizmus*.

More difficult and specialised textbooks came later, eventually incorporating the study of heat engines of all kinds. Of crucial importance in Britain was W. J. M. Rankine, *A Manual of the Steam Engine and other Prime Movers* (1859). For an early French study see M. E. Bataille, *Traité des machines à vapeur* (1847). As late as 1950, books on the steam engine written at this level appeared in the same year in France, Italy and Germany – J. Broch, *La Machine à Vapeur*; M. Medio, *Le Machine termiche*; and A. Uppitz, *Kolbermaschinen*.

For the poetry, see *inter alia* J. Warburg, (Ed.), *The Industrial Muse** (1958) and M. Vicinus's book with the same title (1974); E. M. Grant, *French Poetry and Modern Industry* (1927); and S. Carr (Ed.), *The Poetry of Railways* (1978).

CHAPTER 4

Reading for this chapter should begin with Trevithick and high-pressure steam. See, in particular, H. W. Dickinson and A. Tirley, *Richard Trevithick, the Engineer and the Man** (1934). See also the English translation of the important German study by E. Alban, *The High Pressure Steam Engine* (1847). L. T. C. Rolt's *George and Robert Stephenson: the Railway Revolution** (1960) is indispensable.

So much has been written on railways that even a brief bibliography would fill a book. See, in particular, J. Simmons, *The Railways of Britain** (2nd edn., 1968); O. S. Nock, *Railways Then and Now, A World History* (1975), *Steam Railways in Retrospect* (1966) and the six volumes of his *Railways of the World in Colour* (1972-74);

Hamilton Ellis, *British Railway History* (1954);
M. Robbins, *The Railway Age** (1962); H.
Perkin, *The Age of the Railway** (1970); and
D. Ball, *America's Colourful Railways* (New York,
1978). See also the chapter on Transport by
L. Girard in the *Cambridge Economic History of
Europe* (1965).

Nineteenth-century studies include D.
Lardner, *Railway Economy* (1850); F. S. Williams,
Our Iron Roads (1852); C. E. Stretton, *Safe
Railway Working* (1887), which went through
three editions during the next six years; and
J. Pangborn, *The World's Rail Ways* (1894).

R. L. Stevenson, *Across the Plains* (1892);
L. Beebe and C. Clegg *Hear the Train Blow*
(1952); and Dee Brown, *Hear that Lonesome
Whistle Blow* (1977) refer fully to the folklore.

Steam on the rivers is a major American
theme with monographs on almost every river
and one major general book J. H. Flexner,
*Steam Boats Come True** (1944), which includes
a good bibliography. See also G. H. Preble,
*A Chronological History of the Origin and
Development of Steam Navigation* (1895); S. C.
Gilfillan, *Inventing the Ship* (Chicago, 1935);
H. P. Spratt, *The Birth of the Steamboat* (1959);
G. Body, *British Paddle Steamers* (1971); and
P. W. M. Griffin, *Paddlesteamers* (1968).

More detailed studies of scattered places
include R. E. Lingenfelther, *Steamboats on the
Colorado River* (Arizona, 1978); A. McQueen,
Clyde River Steamers (1923); A. Reid, *Paddle
Wheels on the Wanganu* (Auckland, 1967); and
W. L. S. Eifart, *Delta Queen, The Story of a Steam
Boat* (1960).

For steamships see *inter alia* B. W. Bathe,
Steamships (1969); R. T. Rowland, *Steam at Sea**
(1970); H. P. Spratt, *Transatlantic Paddle-
steamers* (1957); D. B. Tyler, *Steam Conquers
the Atlantic* (1939); and E. A. Wiltsee, *Gold
Rush Steamers of the Pacific* (San Francisco, 1958).

There is much of interest in G. Chen, *Tseng
Kuo Fan, Pioneer Promoter of the Steamship in China*
(1935).

CHAPTER 5

The indispensable introduction to this chapter
is D. S. Landes, *The Unbound Prometheus** (1969)
which covers developments in European
countries comparatively. For a contemporary
statement, see G. F. Kolb, *The Condition of
Nations . . . with Comparative Tables of Universal
Statistics* (1880). See also W. O. Henderson,
The State and the Industrial Revolution in Prussia
(1958); G. C. Allen, *A Short Economic History of
Modern Japan* (1965); and D. H. Buchanan, *The
Growth of Capitalist Enterprise in India* (1966).

For the spread of knowledge of the steam
engine and the extension of its scope, see
J. Tann, and M. J. Brechin, 'The International
Diffusion of the Watt Engine' in the *Economic
History Review*, Vol. 31 (1978).

For the United States, see C. W. Fussell,
*Early Stationary Steam Engines in America: a Study
in the Migration of a Technology* (Washington,
1969).

For France, see J. Payen, *Capital et Machines
à vapeur au XVIIIe siècle* (Paris, 1969); Sir John
Clapham, *The Economic Development of France and
Germany* (1921) and A. Dunham, *The Industrial
Revolution in France* (1955). A course of lectures
delivered by J. Hirsch of the Ecole Nationale des
Ponts et Chaussées on steam engines and
locomotives was published in Paris in 1898.

For the changes in the use of steam see O.
Reynolds, *The Uses of Steam* (1887); C. A.
Augustus, *Steam Using; or Steam Engine Practice*
(Chicago, 1885); G. W. Sutcliffe, *Steam Power

and Mill Work* (1895); C. F. Hirschfield and
T. C. Ulbricht, *Steam Power* (New York, 1916);
and G. Watkins, *The Stationary Steam Engine*
(1968) and *The Steam Engine in Industry** (1978).

The British War Office produced in 1911
'Notes and Memoranda on the Management of
Steam Engines, Boilers, Gas and Oil Engines'.

For boiler explosions see E. B. Marten,
Records of Steam Boiler Explosions (1872); W. H.
Chaloner, *Vulcan, The History of One Hundred
Years of Engineering and Insurance* (1959) and
P. W. J. Bartripp, 'The State and the Steam
Boiler' in the *International Review of Social
History*, Vol. XXV (1980).

For different types of steam engines and their
design and function, see J. A. Ewing, *The
Steam Engine and other Heat Engines** (1894);
W. Norris and B. H. Morgan, *High Speed Steam
Engines* (1900); W. I. Tennant, *The Compound
Engine* (1905); J. Stumpff, *The Unaflow Engine*
(1912); and T. W. Corbin, *Modern Engines*
(1918). See also A. H. Zetban, *Steam Power
Plants* (1952) and P. A. Gaffert, *Steam Power
Stations* (1952).

Cornish engines are dealt with in W. Pole,
A Treatise on the Cornish Pumping Engine (1844)
and D. B. Barton, *The Cornish Beam Engine*
(1967).

The claims of Corliss engines were advocated
powerfully in the Company's 1857 pamphlet,
published in Providence, *The Steam Engine as it
was, and as it is.*

The most useful account of the steam turbine
is Sir Charles Parsons's own Rede lecture of
1911 *The Steam Turbine*. See also R. H. Parsons,
*The Steam Turbine and Other Inventions of Sir
Charles Parsons* (1942); G. Stoney, 'Lectures on
Steam Turbines' in *The Journal of the Society of
Arts* (1909); and A. Stodola, *Steam and Gas
Turbines* (1927).

For steam on the farm, see M. Williams,
Steam Power in Agriculture (1977); H. Bonnett,
Farming with Steam (1974); R. M. Wik, *Steam
Power on the American Farm* (Philadelphia, 1933);
D. C. Jennings, *Days of Steam and Glory* (North
Plains, South Dakota, 1968); and J. Stewart,
Steam Engines on Sugar Plantations (New York,
1867).

CHAPTER 6

Alternative technologies to steam have been the
subject of many books and articles, and the
history of electricity (and electronics), for
example, has attracted at least as much
attention as the history of steam. See S. B.
Goslin, *The Relative Advantages of Wind, Water
and Steam as Motive Powers* (1881); A. Witz,
*Rendement comparé des machines à vapeur et des
machines à gaz* (Paris, 1902); R. Kennedy, *The
Book of Modern Engines: A Practical Work on
Prime Movers and the Transmission of Power:
Steam, Electricity, Water, Gas and Hot Air* (1912);
J. W. Weir, *Modern Power Generation; Steam,
Electric and Internal Combustion; and their
Application to Present-day Requirements* (1908); and
A. Dow, *The Production of Electricity by Steam
Power* (South Bethlehem, Pennsylvania, 1917).
One of the first books to deal with steam and
nuclear power was Q. Wootton, *Steam Cycles for
Nuclear Power Plant* (1958). D. Ross's *Energy from
the Waves* (1979) bears as its breathless sub-
title 'the First-Ever Book on a Revolution in
Technology'.

H. I. Sharlin, *The Making of the Electrical
Age* (1963) is a general introduction. See also
P. Dunsheath, *A History of Electrical Engineering*
(1962). There is much of interest in J. Benthall's
provocative *The Body Electric* (1976).

For the effects of the energy crisis of the
1970s on thinking, see *inter alia* the Committee
for Science and Society, *Deciding about Energy
Policy* (1979); and W. Bach, W. Manstard,
W. H. Matthews and Harrison Brown (Eds.),
Renewable Energy Prospects (1980).

The best introduction to the vintage steam
themes of this chapter is to follow the successive
editions of *The Railway Enthusiasts' Handbook*,*
edited by Geoffrey Body and published by
David and Charles, Newton Abbott, the first
edition of which appeared in 1968/69. One
periodical out of many, the *Railway Magazine*, is
said 'every month' to recreate 'the sight and
smell of steam'. The periodicals are international
in scope and include the *European Railways
Magazine*, the *International Railway Journal*, and
Railway Gazette International. There are also many
periodicals (and societies) devoted to model
railway building.

Books include B. J. Finch, *A Rally of Traction
Engines* (1969); D. Hagan, *Indian Summer of
Steam* (1980); G. Heiron and E. Treacy, *Steam's
Indian Summer* (1979); P. B. Whitehouse,
Preserved Steam in Britain (1979); D. C. Rodgers,
South African Steam Today (1980); D. P. Morgan,
Canadian Steam (1961); Colonel H. C. B. Rogers,
Transition from Steam (1980); Colin Garratt,
Iron Dinosaurs (1976); and Paul Theroux, *The
Railway Bazaar*. See also L. C. Crane, *Long Live
the Delta Queen* (New York, 1973). There are
many books on industrial archaeology. See,
in particular, R. H. Buchanan and G. Watkins,
*The Industrial Archaeology of the Stationary Steam
Engine* (1976).

The sounds of 'by-gone steam' are recaptured
on a number of records by Argo (and others):
they include not only *The Power of Steam* but
Steam in the Worth Valley.

PLACES TO SEE STEAM ENGINES

This list of places where steam engines may be seen is an approximate guide, not an exhaustive gazetteer. The publishers unfortunately cannot accept liability either for its accuracy or for the accessibility of the sites mentioned.

Most detail is given in the United Kingdom listing because the British steam engine 'population density' is much the highest in the world. To aid clarity, England is divided into four areas; Scotland, Wales, and Ireland with the Isle of Man are treated as a further three.

A coding is used to indicate the approximate status of each location and to highlight places where live steam may be enjoyed. But it cannot cover all the permutations of seasonal, weekend and weekday openings which vary from year to year. A telephone call (using Directory Enquiries) to discover the current visiting arrangements should be made before starting a long journey to a site.

Engines mentioned are only the principal engines to be seen on a site; small auxiliary engines are excluded. Also deliberately excluded from the list are modern high-speed all-enclosed reciprocating engines, most steam turbines and engines at establishments where access is restricted.

KEY

▲▲▲

Museum or railway open to the public at weekends, regular steamings

▲▲

Museum or railway open to the public at weekends, occasional steamings

▲

Museum open to the public at normal hours, some engines steamed regularly

△

Museum open to the public at normal hours, no engines in steam

●●

Preserved, occasional public openings

●

Working commercially under steam; stationary engines need prior permission to view

○

Preserved privately, not under steam; apply to company or authority.

SOUTH AND EAST ENGLAND

Berkshire, Buckinghamshire, Cambridgeshire, Essex, Hampshire, Hertfordshire, Isle of Wight, Kent, London, Middlesex, Norfolk, Northamptonshire, Oxfordshire, Suffolk, Surrey, Sussex

Alresford, Hants: Mid-Hants Railway, Alresford Sta. (locos, depot at Ropley)
Brighton Engineerium, Sussex: Goldstone Pumping Sta., Hove (beam & other engines, traction engines, models)
Diss, Norfolk: Bressingham Steam Museum, Bressingham Hall (stationary, railway & traction engines)
Isle of Wight: I.O.W. Steam Railway, Haven Street Sta. (locos)
Leighton Buzzard, Beds: Leighton Buzzard Narrow Gauge Railway, Pages Park (locos)
Liphook, Hants: Hollycombe House Steam Museum (traction & fairground engines)
London: Kew Bridge Pumping Sta., Green Dragon La., Brentford (rare Cornish beam engines; other beam, stationary & traction engines); *see also entries below*
New Romney, Kent: Romney, Hythe & Dymchurch Railway (15 in.-gauge locos)
Peterborough, Northants: Nene Valley Railway, Wansford Sta. (U.K. & Continental locos)
Sheffield Park, Sussex: Bluebell Railway, Sheffield Park Sta. (locos)
Sittingbourne, Kent: Sittingbourne & Kemsley Light Railway, Milton Road (narrow-gauge locos)
Tenterden, Kent: Kent & East Sussex Railway (locos; depot at Rolvenden)

Blenheim, Oxon: Combe Sawmill, Blenheim Estate (1852 beam engine)
Didcot, Oxon: Didcot Railway Centre, Didcot Sta. (mainly GWR locos)
Halstead, Essex: Colne Valley Railway, Castle Hedingham Sta. (locos)
Portsmouth, Hants: Eastney Sewage Pumping Sta. (2 1884 beam engines)
Quainton, Aylesbury, Bucks: Quainton Railway Society, Quainton Sta. (locos)
Sheringham, Norfolk: North Norfolk Railway, Sheringham (locos)
Thursford, Norfolk: T. Cushing's Museum (traction engines & fair organs)

London: *HMS Belfast*, opposite Tower of London, S.E.1. (marine turbine installation); London Transport Museum, Covent Garden, W.C.1. (locos & vehicles); National Maritime Museum, Greenwich (marine engines & models); Science Museum, Exhibition Rd., S.W.7. (major national collection); *see also entries above & below*
Stretham, Ely, Cambridgeshire: Stretham beam engine & scoop wheel

Cambridge: Cheddars Lane Pumping Sta. (2 rare Hathorn Davey differential pumping engines)

Croydon, Surrey: Waddon Pumping Sta. (2 horizontal engines; Thames Water Authority)
High Wycombe, Bucks: Thomas Glenister Co., furniture makers, Temple End (horizontal compound and overtype engines)
Hook Norton, Oxon: Hook Norton Brewery Co. Ltd. (horizontal engine)
Horsham, Sussex: King & Barnes Ltd., brewery (horizontal engine)
London: Young & Co. Ltd., brewery, Wandsworth High Street, S.W.18. (2 small beam engines); *see also entries above & below*

Ashford, Kent: Henwood Pumping Sta. (2 beam engines, 1870–80; Southern Water Authority)
Brockham, Surrey: Narrow-gauge Railway Museum (locos)
Cheshunt, Herts: Turnford Pumping Sta. (rare 1845 side-lever engine; Thames Water Authority)
Erith, Kent: Crossness Sewage Pumping Sta. (4 large beam engines; Thames Water Authority)
Folkestone, Kent: Upper Cherry Gardens Waterworks (2 1889 Worthington triple non-rotative engines, plus others; Southern Water Authority)
Hanworth, Middx: Kempton Park Pumping Sta. (2 large inverted vertical engines; Thames Water Authority)
Hastings, Sussex: Brede Pumping Sta. (2 inverted vertical engines; Southern Water Authority)
London: Tower Bridge, S.E.1. (horizontal steam-hydraulic pumping engine; apply to City of London) West Ham Sewage Pumping Sta., E.15. (2 beam engines; Thames Water Authority); *see also entries above*
Lound, Suffolk: Lound Pumping Sta. (2 1854 grasshopper beam engines; Anglian Water Authority)
Staines, Middx: Littleton Pumping Sta. (rare Uniflow engine; Thames Water Authority)

SOUTH & WEST ENGLAND

Avon, Cornwall, Devon, Dorset, Gloucestershire, Herefordshire, Somerset, Wiltshire, Worcestershire

▲▲▲

Buckfastleigh, Devon: Dart Valley Railway, Buckfastleigh Sta. (main-line & industrial locos)
Minehead, Somerset: East Somerset Railway, Minehead Sta. (a few locos)
Paignton, Devon: Torbay Steam Railway, Paignton Sta. (main-line locos)

▲▲

Aschurch, Glos: Dowty Railway Preservation Centre (std. and narrow-gauge ind. locos)
Great Bedwyn, Wilts: Crofton Pumping Sta. (1812 & 1846 small Cornish beam engines)
Hereford: Broomy Hill Waterworks Museum, Broomy Hill (triple expansion pumping & other engines) Bulmer's Railway Centre, Whitecross Rd. (main-line & ind. locos)
Lydney, Glos: Dean Forest Railway, Norchard Steam Centre, New Mills (a few locos)
Redruth, Cornwall: Mawla Well Farm Museum, Mawla (a few traction engines)
Shepton Mallet, Somerset: East Somerset Railway, Cranmore (main-line locos)
Swanage, Dorset: Swanage Railway Centre, Swanage Sta. (2 main-line locos)
Washford, Somerset: Somerset & Dorset Railway Museum, Washford Sta. (mainly ind. locos)
Westonzoyland, Somerset: Westonzoyland Pumping Sta. (1861 vertical engine)

▲

Exeter, Devon: Maritime Museum, City Basin (c. 1840 drag-boat dredger & steam tug)
East Budleigh, Devon: Bicton Countryside Museum (narrow-gauge loco, traction engines)

△

Bristol, Avon: *SS Great Britain*, Gasferry Rd. (replica engines to be fitted); Industrial Museum,

Princes Wharf (steam dockside crane, swing-bridge engine); *see also entry below*

Redruth, Cornwall: Tolgus Tin Museum, Gilbert's Coombe (Cornish rotative beam engine to be erected); *see also entries above & below*

St. Austell, Cornwall: Wheal Martyn Clay Industry Museum, Carthew (2 ind. locos)

St. Just, Cornwall: Geevor Mine, Pendeen (1840 beam winding engine at Levant, & horizontal winding engine)

Swindon, Wilts: Great Western Railway Museum, Farringdon Rd. (G.W. locos)

Taunton, Somerset: Taunton Museum (beam engine)

Wendron, Helston, Cornwall: Poldark Mining Museum (stationary, marine & traction engines)

● **Blandford St. Mary, Dorset:** Hall & Woodhouse Ltd., brewery (2 horizontal engines)

Falmouth, Cornwall: Falmouth Docks (3 steam tugs & 3 ind. locos; P. & O. Group, London)

Redruth, Cornwall: Devenish & Co., Redruth Brewery (small horizontal engines); *see also entries above and below*

○ **Blagdon, Somerset:** Blagdon Pumping Sta. (pair of 1902 beam engines; Wessex Water Authority)

Bristol, Avon: Underfall Maintenance Yard (steam-driven machinery; Port of Bristol Authority); *see also entry above*

Dartmouth, Devon: Newcomen Engine House, The Butterwalk (1725 atmospheric engine)

Malvern, Glos: Broomsberrow Pumping Sta., Broomsberrow (horizontal engine; Severn-Trent Water Authority)

Othery, Somerset: Aller Moor Pumping Sta., Burrow Bridge (1869 vertical engine, 3 others nearby; Wessex Water Authority)

Redruth, Cornwall: East Pool Mine, Pool (Cornish beam pumping & winding engines); South Crofty Mine, Pool (1854 Cornish beam pumping engine); *see also entries above*

Street, Somerset: C.&J. Clark, shoe manufacturers (compound mill engine)

St. Austell, Cornwall: Goonvean & Restowrack China Clay Co., Goonvean Clay-works, St. Stephen-in-Brannel (Cornish beam pumping engine); English China Clays, Parkandillick Clayworks, St. Dennis (Cornish beam pumping engine & haulage engine)

Uffculme, Devon: Fox Bros. & Co. Ltd., textile mill, Coldharbour (compound engine)

NORTH & WEST ENGLAND
Cheshire, Cumbria, Derbyshire, Greater Manchester, Lancashire, Merseyside, Shropshire, Staffordshire, West Midlands

▲▲▲ **Bridgnorth, Salop:** Severn Valley Railway (28 main-line locos)

Carnforth, Lancs: Steamtown Railway Museum, Warton Rd. (over 30 main-line, Continental & ind. locos)

Glossop, Manchester: Dinting Railway Centre (main-line & ind. locos)

Ravenglass, Cumbria: Ravenglass & Eskdale Railway (15 in.-gauge locos)

Stoke-on-Trent, Staffs: Foxfield Light Railway, Dilhorne, Blythe Bridge (12 ind. locos); *see also entries below*

Telford, Salop: Ironbridge Gorge Museum, Blist's Hill, Ironbridge (small winding engine & other engines)

Ulverston, Cumbria: Lakeside & Haver-thwaite Railway, Haverthwaite Sta., Newby Bridge (10 locos, mostly ind.)

▲▲ **Birmingham, W. Midlands:** Railway Museum, Warwick Rd., Tyseley (main-line locos); *see also entries below*

Brownhills, W. Midlands: Chasewater Light Railway, Chasewater Pleasure Park (ind. locos)

Kendal, Cumbria: Levens Hall Steam Collection, M6 jct. 36 (steam road vehicles & models)

Leek, Staffs: North Staffordshire Railway, Cheddleton Sta. (3 locos)

Matlock, Derbyshire: Lea Wood Pumping Sta., Cromford Canal (1849 Cornish beam engine)

Millmeece, Staffs: Millmeece Pumping Sta. (large 1914 & 1926 horizontal engines; Severn-Trent Water Authority)

Ripley, Derbyshire: Midland Railway Centre, Butterley Sta. (mostly main-line locos)

Royton, Lancs: Diamond Rope Works, Royton (vertical engine; Northern Mill Engine Society)

Shaw, Lancs: Dee Mill, Cheetham St. (1907 4-cyl. compound engine; Northern Mill Engine Society)

Windermere, Cumbria: Steamboat Museum, Lake Windermere (vintage steam launches)

▲ **Birmingham, W. Midlands:** Museum of Science & Industry, Newhall St. (major collection, inc. Watt 1779 beam engine); *see also entries above & below*

Great Haywood, Staffs: Museum of Stafford-shire Life, Shugborough Hall (ind. locos)

△ **Burton-on-Trent, Staffs:** Bass Museum, Horninglow St. (loco & mill engine); *see also entry below*

Liverpool, Merseyside: County Museum (early loco & steam vehicles)

Lytham, Lancs: Lytham Motive Power Museum, Dock Rd. (locos & road vehicles)

Manchester: North-Western Museum of Science & Industry, Grosvenor St. (replica atmospheric & other engines)

Southport, Merseyside: Steamport Transport Museum, Derby Rd. (10 locos)

Stoke-on-Trent, Staffs: Chatterley-Whitfield Colliery Museum, Tunstall (large 1914 winding engine); *see also entries above & below*

Wirksworth, Derbyshire: Middleton Top Engine (1829 double-beam winding engine)

● **Bolton, Lancs:** Atlas Forge, Bridgeman St. (2 vertical rolling mill engines); *see also entry below*

Briercliffe, Lancs: Queen Street Mill, Harle Syke (1895 compound engine)

Burslem, Staffs: Dalehall Works, Stubb's St., Middleport (small mill engine)

Church, Burnley, Lancs: India & Primrose Mills, Bridge St. (1884 compound engine)

Hartington, Derbyshire: DSF Refractories, Friden Brickworks (1902 mill engine)

Killamarsh, Derbyshire: Westthorpe Colliery (1924 winding engine; National Coal Board)

Leek, Staffs: Cheddleton Flint Mill (horizontal engine)

Padiham, Lancs: Padiham Room & Power Co., Jubilee Mill (1888 compound engine)

Pleasley, Derbyshire: Pleasley Colliery (large 1902 & 1924 winding engines; N.C.B.)

St. Helens, Lancs: Sutton Manor Colliery (several large engines; N.C.B.)

Uttoxeter, Derbyshire: Klondyke Mill, Draycott-in-the-Clay, Sudbury (traction engines & narrow-gauge locos)

Whitehaven, Cumbria: Haig Colliery (winding & haulage engines; N.C.B.)

Workington, Cumbria: Moss Bay Steelworks, Harrington Rd. (horizontal rolling mill engine; British Steel Corporation)

○ **Bamford, Derbyshire:** Carbolite Co. Ltd., Bamford Mill (1907 mill engine)

Birkenhead, Merseyside: Shore Rd. Pumping Sta. (1884 non-rotative grasshopper beam engine; British Rail)

Birmingham, W. Midlands: Grazebrook Blowing Engine, A38(M) roundabout, Aston Expressway (on public view); *see also entries above*

Bolton, Lancs: Red Bridge Mills, Ainsworth (Vee-type mill engine); Northern Mill Engine Society collection, Atlas Mills, Chorley Old Rd. (17 engines, mostly dismantled); *see also entry above*

Burton-on-Trent, Staffs: Clay Mills Sewage Pumping Sta. (2 1885 beam & many small engines); *see also entry above*

Ellesmere Port, Cheshire: North Western Museum of Inland Navigation, Upper Engine House (hydraulic pumping engines, 1873 & c. 1910); Manchester Ship Canal Co., Lower Engine House (hydraulic pumping engines 1866)

Leigh, Lancs: Leigh Spinners No. 2 Mill, Park La. (large 1925 compound engine)

Lichfield, Staffs: Sandfields Pumping Sta. (1873 Cornish beam engine; Severn-Trent Water Authority)

Shaw, Lancs: Fern Mill (1884 4-cyl. compound engine; N.M.E.S.)

Shrewsbury, Salop: Coleham Head Sewage Pumping Sta., Longden, Coleham (pair of 1897 beam engines)

Silloth, Cumbria: Carr's Flour Mills, Solway Mills (Belgian-built horizontal engine)

Stoke-on-Trent, Staffs: Etruscan Bone & Stone Mill, Etruria (1858 beam engine; City of Stoke-on-Trent); *see also entries above*

Tyldesley, Lancs: Astley Green Colliery (1912 4-cyl. winding engine; local authority)

Wigan, Lancs: Trencherfield Mill, Wallgate (1907 4-cyl. engine; Jeffreys Miller & Co., Leyland Mills)

Wombourn, Staffs: Bratch Pumping Sta. (2 inverted vertical engines; Severn-Trent Water Authority)

NORTH & EAST ENGLAND
Durham, Humberside, Leicestershire, Lincoln-shire, Nottinghamshire, Northumberland, Yorkshire

▲▲▲ **Guisburn, Yorks:** Todber Museum of Steam on A682 (15 traction engines, wagons & rollers)

Keighley, Yorks: Keighley & Worth Valley Railway, Keighley Sta. (30 main-line & ind. locos)

Leeds, Yorks: Middleton Railway, Middleton (ind. locos)

Loughborough, Leics: Great Central Railway, Loughborough Central Sta. (17 main-line & ind. locos)

Market Bosworth, Leics: Market Bosworth Light Railway, depot at Shackerstone Sta. (10 mostly ind. locos)

Moira, Leics: Rawdon Colliery (1868 winding engine; N.C.B.)

Nottingham: Wollaton Park Industrial Museum (beam engine)

Pickering, Yorks: North Yorkshire Moors Railway, depot at Grosmont Sta. (19 main-line & ind. locos)

Skipton, Yorks: Yorkshire Dales Railway, Embsay Sta. (15 mainly ind. locos)

▲▲

Hucknall, Notts: Papplewick Pumping Sta. (pair of 1884 beam engines)

Leicester, Leics: Museum of Technology, Abbey Pumping Sta., Corporation Rd. (four 1890–91 compound beam engines, steam shovel & other engines)

Sunderland, Tyne & Wear: Ryhope Pumping Sta., Ryhope (pair of 1868 beam engines)

Tattershall, Lincs: Dogdyke Pumping Sta. (1856 beam engine & scoop wheel)

△

Bradford, Yorks: Industrial Museum, Moorside Mills, Moorside Rd. (ind. loco & small engines)

Hull, Humberside: Transport Museum (steam vehicles)

Newcastle-on-Tyne: Museum of Science & Engineering, Exhibition Park (beam engines, loco, marine & other engines)

Sheffield, Yorks: Abbeydale Industrial Hamlet (horizontal engine)

Stanley, Co. Durham: Beamish Open Air Museum (overcrank winding engine, steam shovel, locos, other engines)

Washington, Tyne & Wear: Washington F. Colliery (1888 winding engine; local authority)

York: National Railway Museum (major collection of main-line locos)

●

Hucknall, Notts: Linby Colliery, Linby (2 1922 winding engines; N.C.B.)

Mansfield Woodhouse, Notts: Sherwood Colliery (2 1903 winding engines; N.C.B.)

Otley, Yorks: Wm. Barker & Sons, tannery, Cross St. (1907 mill engine)

Seaham, Tyne & Wear: Seaham Harbour (steam tug)

Sheffield, Yorks: Orgreave Colliery, Handsworth (1928 winding engine; N.C.B.)

Thurnscoe, Yorks: Hicketon Colliery (large 1912 winding engine; N.C.B.)

○

Bradford, Yorks: New Lane Mills, Laisterdyke (compound Uniflow engine)

Bulwell, Notts: Bestwood Colliery, Bestwood (vertical winding engine; local authority)

Darlington, Yorks: Broken Scar Pumping Sta., Coniscliffe Rd. (1904 beam engine; Yorkshire Water Authority); North Road Railway Museum, North Road Sta. (6 locos 1825–1909)

Holmfirth, Yorks: Washpit Mills (compound engine)

Hoyland Nether, Yorks: Elsecar Colliery (1795 atmospheric pumping engine; N.C.B.)

Hull, Humberside: Springhead Pumping Sta.

(1876 Cornish beam engine)

Kellington, Yorks: Roall Pumping Sta. (1891 compound beam engine)

Kirkby-in-Ashfield, Notts: Bentinck Colliery (1915 winding engine; N.C.B.)

Middlestown, Yorks: Caphouse Colliery, Overton (1876 winding engine; N.C.B.)

Ossett, Yorks: Oakliffe Engineering Co., Runtlings Mill (1908 compound engine)

Haxey, Lincs: Owston Ferry Pumping Sta. (1904 tandem compound engine)

Shepley, Yorks: Firth Bros. Shepley New Mills (inverted vertical engine)

Spalding, Lincs: Pinchbeck Marsh Pumping Sta. (1933 beam engine & scoop wheel; Anglian Water Authority)

Sunderland, Tyne & Wear: Dalton Pumping Sta., Cold Hesledon (pair of 1879 Cornish beam engines)

Wakefield, Yorks: Walton Colliery, Walton (2 large winding & other engines)

Worksop, Notts: Lound Hall Mining Museum (winding and other engines; N.C.B.)

WALES

▲▲▲

Aberystwyth, Dyfed: Vale of Rheidol Railway (narrow-gauge locos)

Bala, Gwynedd: Bala Lake Railway, Llanuwchllyn Sta. (locos)

Carmarthen, Dyfed: Gwili Railway, Bronwydd Arms Sta. (mostly ind. locos)

Fairbourne, Gwynedd: Fairbourne Railway (15 in.-gauge locos)

Llanberis, Gwynedd: Llanberis Lake Railway (n-g locos); Snowdon Mountain Railway (Swiss-type rack locos)

Porthmadog, Gwynedd: Festiniog Railway, depot at Boston Lodge (n-g locos, inc. 2 Fairlie type)

Tywyn, Gwynedd: Talyllyn Railway, depot in Tywyn (n-g locos); *see also entry below*

Welshpool, Powis: Welshpool & Llanfair Light Railway, depot at Llanfair Caereinion (n-g locos, inc. 1 Mallet type & 2 foreign)

▲▲

Caerphilly, Glamorgan: Caerphilly Railway Society, Harold Wilson Ind. Estate, Van Road (locos)

Cardiff, Glamorgan: Welsh Industrial & Maritime Museum, Bute Dock (steam tug & stationary engines)

Swansea, W. Glamorgan: Industrial & Maritime Museum, South Dock (steam tug & ind. locos)

△

Bangor, Gwynedd: Penrhyn Castle Ind. Railway Museum (n-g locos)

Blaenau Ffestiniog, Gwynedd: Ffestiniog Mountain Tourist Centre (n-g locos)

Cardiff, Glamorgan: National Museum of Wales (a few stationary engines)

Llangollen, Clwyd: Llangollen Railway Society (ind. locos)

Tywyn, Gwynedd: Narrow-gauge Museum, Tywyn Wharf (n-g locos); *see also entry above*

○

Barry, S. Glamorgan: Woodham Bros., Barry Dock (more than 100 derelict main-line locos)

Crynant, W. Glamorgan: Blaenant Colliery (large winding engine; N.C.B.)

Cwm, Gwent: Marine Colliery (derelict pumping & haulage engines c. 1895, loco; N.C.B.)

Kidwelly, Dyfed: former Kidwelly Tinplate Works (derelict large rolling mill engines; local authority)

Mountain Ash, Glamorgan: Abergorki Colliery (fan engine, Nat. Mus. of Wales)

Nantlle, Gwynedd: former Dorothea Slate Quarries (1904 Cornish beam engine)

New Tredegar, Glamorgan: Elliott Colliery (compound winding engine; Nat. Mus. of Wales)

Pontypool, Gwent: Glyn Pit, off Crumlin Rd. (derelict beam & overcrank winding engines, c. 1845)

Pontypridd, Glamorgan: Tymawr Colliery (1875 winding engine; N.C.B.)

SCOTLAND

▲▲▲

Aviemore, Highland: Spey Valley Railway, Boat of Garten Sta. (main-line & ind. locos)

Glasgow, Strathclyde: *PS Waverley*, Waverley Terminal (summer passenger service); *see also entry below*

Loch Katrine, Central: *SS Sir Walter Scott*, (1899 steamer, summer passenger service)

Loch Lomond, Strathclyde: *PS Maid of the Loch*, Balloch Pier (summer passenger service, also rare slipway engine)

Lochgilphead, Highland: *SS Vic 32*, Clyde puffer vessel, based at Crinan (holiday cruises on Caledonian Canal)

▲▲

Edinburgh, Lothian: Prestongrange Mining Museum, Morrisons Haven, Prestonpans (1853 Cornish beam & other engines, & ind. locos); *see also entry above*

Falkirk, Central: Scottish Railway Preservation Society, Wallace St. (main-line & ind. locos)

Kilconquhar, Fife: Lochty Private Railway, Lochty Farm, Balbuthie (a few locos)

△

Edinburgh, Lothian: Royal Scottish Museum, Chambers St. (1786 beam engine, 1813 loco, other engines, models); *see also entry above*

Glasgow, Strathclyde: Museum of Transport, 25 Albert Rd. S1 (main-line locos, road steam vehicles, marine engines, models); *see also entry above*

Renfrew, Strathclyde: Clyde Port Authority, Renfrew Ferry (spare ferry & other steam vessels); *see also entry below*

●

Auchterarder, Tayside: Glenruthven Mills (1873 horizontal engine)

Dumbarton, Strathclyde: Town Centre (1824 side-lever engine, on permanent view)

Renfrew, Strathclyde: Ferry Gardens (1851 side-lever engine on permanent view); *see also entry above*

NORTHERN IRELAND, EIRE, ISLE OF MAN

▲▲▲

Antrim, Country Antrim, N.I.: Shane's Castle Railway (narrow-gauge locos)

▲▲
Stradbally, County Laois, Eire: Irish Steam Preservation Society, The Green (n-g locos & traction engines)
Whitehead, County Antrim, N.I.: Railway Preservation Society of Ireland, Whitehead Excursion Sta. (9 locos)

△
Armagh, County Armagh, N.I.: County Museum, The Mall East (road & rail transport)
Belfast, N.I.: Belfast Transport Museum, Witham St. (main-line locos)
Dublin, Eire: Guinness Museum (1882 overtype n-g locos)
Holywood, County Down, N.I.: Ulster Folk & Transport Museum, Cultra Manor, 8 miles from Belfast (road & rail transport)
Port Erin, Isle of Man: Port Erin Railway Museum, Port Erin Sta. (served by IOMR: n-g locos)

PLACES TO SEE STEAM OUTSIDE U.K.

ARGENTINA*
Buenos Aires: Port Area – active steam tugs
Quilmes: Museum of the History of Transport

AUSTRALIA
Adelaide: Railway Museum
Brisbane: Redbank Railway Museum
Fremantle: Maritime & Historical Museum
Goulbourne: Museum of Historic Engines
Mannum: Marion Paddlewheel Museum
Melbourne: Railway Museum; Science Museum of Victoria
Parramatta: Steam Tram & Railway Museum
Sydney: Museum of Applied Arts & Sciences; Rail Transport Museum

AUSTRIA
Bad Wimsbach: Transport Museum
Tyrol: Zillertahlbahn n-g line
Vienna: Austrian Railway Museum; Museum of Industry & Technology

BELGIUM
Antwerp: National Maritime Museum; Open Air Maritime Museum
Brussels: Belgian Railway Museum; Tramway Museum
Phillippeville (Ardennes): Viroin Valley Tramway

BRAZIL
Campinas: Railway Museum
Porto Velho: Madeira-Mamoré Railway Museum

CANADA
Dawson City: SS *Keno* Museum
Hamilton: Pump House & Steam Museum
Montreal: Canada Railways Museum
Ottawa: National Museum of Science & Technology
Quebec: St. Constant Railway Museum
Toronto: Ontario Science Centre
Whitehouse: SS *Klondyke* Museum

CHINA*
Tientsin: People's Hall of Science

CZECHOSLOVAKIA*
Kosice: Technical Museum
Prague: National Technical Museum;
Danube steamer services

DENMARK
Helsingor: Danish Technical Museum
Maribo: Maribo-Bandholm Steam Railway
Odense: Railway Museum

EGYPT
Cairo: State Railways Museum

FINLAND
Hyvinka: Railway Museum
Savonlinna: SS *Salama*

FRANCE
Compiègne: Museum of Road Transport
Mulhouse: Railway Museum
Paris: National Technical Museum (Museum of Arts and Trades)
Pithiviers: Transport Museum
Uzes: Museon di Rodo

E. GERMANY*
Dresden: Transport Museum
Wernigerode: Harzquer bahn

W. GERMANY
Achern: Ottenhöfen bahn
Berlin: Transport Museum
Cologne & Mainz: Rhine paddle-steamer services
Frankfurt: Model Railway Museum
Koblenz: Rhine Museum
Munich: Deutsches Museum
Nuremberg: Transport Museum

HUNGARY
Budapest: Hungarian Transport Museum

INDIA*
Delhi: Railway Museum

ITALY
Como: Lake steamers
Milan: Leonardo da Vinci Museum of Science and Industry

JAPAN
Tokyo: Museum of Transport & Technology

NETHERLANDS
Hoorn: Medemblik Steam Tram
Kamperzeedijk: Mastenbroek Steam Engine Museum
Utrecht: Netherlands Railway Museum
Vijfhuizen: Cruquius Cornish Beam Engine Museum
Weesp: Netherlands Tramways Museum

NEW ZEALAND
Auckland: Museum of Transport and Technology
Christchurch: Ferrymead Trust

NORWAY
Hamar: Railway Museum
Løke Møsa: PS *Skibladner*
Oslo: Norwegian Science and Industry Museum; SS *Borovsund*
Søramsund: Hølandsbanen (steam railway)

PERU*

POLAND*
Warsaw: Railway Museum

ROMANIA
Bucharest: Railway Museum

SOUTH AFRICA*
Johannesburg: Crown Gold Mines Museum; South African Railways Museum

SPAIN
Madrid: Railway Museum

SWEDEN
Gavle: Swedish Railways Museum
Stockholm: Steamer services; Technical Museum

SWITZERLAND
Lake Geneva: Paddle-steamer services
Interlaken: Rothorn rack railway
Lake Lucerne: Paddle-steamer services
Lucerne: Swiss Museum of Transport
Montreux: Blonay-Chamby Railway
Lake Zurich: Paddle-steamer services
Winterthur: Swiss Technorama

TURKEY*
Istanbul: Bosphorus ferry steamers

USA
Ashland: Santa Fe Museum
Baltimore: Baltimore & Ohio Transportation Museum
Bellows Falls: Steamtown
Chattanooga: Tennessee Valley Railroad Museum
Corinne: Railroad Museum
Dallas: 'Age of Steam' Museum
Dearborn: Henry Ford Museum
Exton: Thomas Newcomen Memorial Library & Museum
Fall River: Marine Museum
Fort Worth: Pate Museum of Transportation
Green Bay: National Railroad Museum
Houston: Battleship *Texas*
Long Beach: RMS *Queen Mary*
Louisville: Kentucky Railway Museum
Marietta: Steamer *WP Snyder Jnr*
New York: Brownville Railroad Museum; South St. Seaport Museum
Oneonta: National Railroad Museum
St. Louis: National Museum of Transport
Sandy Creek: Rail City
Seattle: Pugat Sound Railway Museum
South Carver (Mass.): Edaville Railroad
Strasburg: Strasburg Railway & Railroad Museum of Pennsylvania
Truckee: Emigrant Trail Museum
Union: Illinois Railway Museum
Washington D.C.: National Museum of History & Technology (Smithsonian Institution)
Wilmington: USS *North Carolina*
Winona: Steamboat Museum
Worthington: Ohio Railway Museum

USSR*
Leningrad: Museum of Railway History

VENEZUELA
Caracas: Museum of Transport

YUGOSLAVIA
Belgrade: Museum of Yugoslav State Railways

ZIMBABWE*
Bulawayo: Zimbabwe Railways Museum

*Parts of the national railway system are steam-operated (1980)

INDEX

ACKNOWLEDGEMENTS

The author and editors wish to extend special thanks to Kenneth Brown, who as consultant gave unstintingly of his time and technical knowledge, offering advice on both the text and the pictures and so materially altering the shape of the book; to David de Haan, Curator of the Elton collection, whose scholarship and helpfulness provided a prop on which we unashamedly leant; to his staff at the Ironbridge Gorge Museum Trust; to Professor Jack Simmons, who gave much authoritative advice on the chapter on Locomotion; and to Susan Hard and Frances McDonald for help throughout.

Grateful thanks are also due to: Rosemary Allan, Beamish North of England Open Air Museum; Harold Bonnett; John Cornwell, Bristol; Chris Coupland, Mary Evans Picture Library; Lord Gibson-Watt of the Wye, Llandrindod Wells; Paul Gunn, London; Susan Norman, Newcastle; Georgina Robins, London; Peter G. Smart, Secretary, The Road Locomotive Society; W. J. Webb, British Fairground Society; Marjorie Willis, BBC Hulton Picture Library.

PICTURE CREDITS

Sources for pictures in this book are separated from left to right by commas, from top to bottom by dashes.

Jacket Front Cover – designed by Roy Williams, cast by Newton Replicas and photographed by Francis Lumley. Jacket Back Cover – Colin Garratt, Photo. Science Museum, London – Crown Copyright. Science Museum, London – Crown Copyright. Science Museum, London, Ironbridge Gorge Museum Trust. First endpaper – University of Reading, Institute of Agricultural History and Museum of English Rural Life. Last endpaper – Ironbridge Gorge Museum Trust. Frontispiece – BBC Hulton Picture Library. Contents Page – University of Reading, Institute of Agricultural History and Museum of English Rural Life. 6 – Ironbridge Gorge Museum Trust. 8 – Ironbridge Gorge Museum Trust. 9 – Agricola, *De Re Metallica*, John Cornwell. 10 – Ironbridge Gorge Museum Trust. 11 – BBC Hulton Picture Library. 12 – University of Reading, Institute of Agricultural History and Museum of English Rural Life. 13 – Royal Institution of Cornwall. 14 – Illustrated London News Picture Library. 15 – R. K. Evans. 16 – Frank D. Woodall. 18 – Hero, *Opera*. 20 – Photo. Science Museum, London. 21 – R. H. Thurston, *A History of the Growth of the Steam-Engine*. 22 – Photo. Science Museum, London. 23 – BBC Hulton Picture Library. 25 – The Walker Art Gallery, Liverpool – Salford Art Gallery and Museum. 26, 27 – Ironbridge Gorge Museum Trust. 28 – Colin Bowden. 31 – Ironbridge Gorge Museum Trust. 33 – Worcester College, Oxford. 34 – R. H. Thurston, *A History of the Growth of the Steam-Engine*. 35 – *Ladies' Diary*. 36 – Ironbridge Gorge Museum Trust and R. H. Thurston, *A History of the Growth of the Steam-Engine*. 38–42 – John Cornwell. 43 – Beamish North of England Open Air Museum. 44 – John Cornwell. 45 – John Cornwell – Beamish North of England Open Air Museum. 46 – Beamish North of England Open Air Museum. 47 – John Cornwell. 49 – Ironbridge Gorge Museum Trust. 50 – Peter Newark's Western Americana. 51 – By courtesy of Rt. Hon. Lord Gibson-Watt of the Wye, 52, 53 – Science Museum, London. 54 – Dionysius Lardner, *The Steam Engine* – Ironbridge Gorge Museum Trust. 55 – Dionysius Lardner, *The Steam Engine*. 56 – Ironbridge Gorge Museum Trust – R. L. Galloway, *The Steam Engine and Its Inventors*. 57 – J. Farey, *A Treatise on the Steam Engine* – R. H. Thurston, *A History of the Growth of the Steam-Enginr*. 58–62 – Ironbridge Gorge Museum Trust. 64 – Mary Evans Picture Library – Illustrated London News Picture Library. 65 – *Scientific American*, 1884. 66 – Ironbridge Gorge Museum Trust – *Scientific American*, 1876. 67 – Mary Evans Picture Library. 68 – Peter Newark's Western Americana – *Scientific American*, 1869, Peter Newark's Western Americana. 69 – Peter Newark's Western Americana. 70 – BBC Hulton Picture Library. 72, 73 – University of Reading, Institute of Agricultural History and Museum of English Rural Life. 74, 75 – Ironbridge Gorge Museum Trust. 76 – BBC Hulton Picture Library. 77, 78 – Ironbridge Gorge Museum Trust. 79 – Birmingham Reference Library. 80 – Ivan Belcher. 81 – BBC Hulton Picture Library. 82 – Institution of Royal Engineers, Royal Engineers Corps Library, Chatham – University of Reading, Institute of Agricultural History and Museum of English Rural Life – BBC Hulton Picture Library. 83 – The Tank Museum, Imperial War Museum – Imperial War Museum, Lent to Science Museum, London by G. Cochrane & Co. – National Army Museum. 84 – Local Studies Library, Nottinghamshire County Library. 85 – Popperfoto – British Fairground Society – British Fairground Society. 86 – British Fairground Society. 89 – Mary Evans Picture Library. 91, 92 – Daniel Meadows except 99 top right – Stanley Graham. 104 – Royal Institution of Cornwall. 106 – Mary Evans Picture Library. 108 – Photo. Science Museum, London. 109 – Ironbridge Gorge Museum Trust – Beamish North of England Open Air Museum. 111 – Crown Copyright. 112 – Leicestershire Museums, Art Galleries and Records Services. 113 – Popperfoto. 114 – Ironbridge Gorge Museum Trust. 115 – Ironbridge Gorge Museum Trust – Mary Evans Picture Library, Ironbridge Gorge Museum Trust. 116 – Union Pacific Railroad Museum Collection. 117 – The Oakland Museum History Department. 118 – Peter Newark's Western Americana. 119 – Ironbridge Gorge Museum Trust. 120 – Dr. L. A. Nixon. 121 – BBC Hulton Picture Library – Ironbridge Gorge Museum Trust. 122 – Ironbridge Gorge Museum Trust. 123 – Peter Newark's Western Americana. 124 – Ironbridge Gorge Museum Trust. 125 – Novosti Press Agency. 126 – R. H. Thurston, *A History of the Growth of the Steam-Engine*. 128 – Bennet Woodcroft, *The Origin and Progress of Steam Navigation*. 129 – National Gallery of Art, Washington. Gift of Mrs. Huttleston Rogers – Ironbridge Gorge Museum Trust. 130, 131 – Peter Newark's Western Americana. 132 – Mary Evans Picture Library – Crown Copyright. Science Museum, London. 133 – BBC Hulton Picture Library. 134 – Brunel University Library. 135 – BBC Hulton Picture Library. 136 – Popperfoto – Waterways Museum, British Waterways Board. 138–147 – BBC Hulton Picture Library. 148 – University of Reading, Institute of Agricultural History and Museum of English Rural Life. 150 – Glenbow Museum Archives, Alberta. 151 – BBC Hulton Picture Library. 152 – Peter Newark's Western Americana. 154 – Road Locomotive Society. 155 – Sir Benjamin Stone Collection, Birmingham Reference Library. 156 – Cruquius Preservation Society. 157 – Papplewick Pumping Station Trust. 158 – Museum of London – BBC Hulton Picture Library. 160 – Ray Hooley, BBC Hulton Picture Library – BBC Hulton Picture Library, R. Ackrill Ltd., Harrogate – University of Reading, Institute of Agricultural History and Museum of English Rural Life. 161 – University of Reading, Institute of Agricultural History and Museum of English Rural Life – Dr. L. A. Nixon – Road Locomotive Society. 162 – Ironbridge Gorge Museum Trust – Ironbridge Gorge Museum Trust – R. H. Thurston, *A History of the Growth of the Steam-Engine*. 163 – Ironbridge Gorge Museum Trust. 164 – British Engine Insurance Ltd. 165 – Peter Newark's Western Americana – University of Reading, Institute of Agricultural History and Museum of English Rural Life. 166 – Ironbridge Gorge Museum Trust. 169 – BBC Hulton Picture Library. 170 – Ironbridge Gorge Museum Trust, The Welbeck Gallery, The Welbeck Gallery. 171 – Ironbridge Gorge Museum Trust. 172 – Ironbridge Gorge Museum Trust. 175 – D. A. Rayner. 176 – Robey & Co. Ltd. 177 – National Motor Museum, Beaulieu. 178 – Colin Bowden. 179 – Beamish North of England Open Air Museum, Kenneth Brown. 181 – Ivan Belcher. 182 – Colin Garratt. 183 – Colin Bowden. 184 – Dr. L. A. Nixon – Colin Bowden. 185 – Georgetown Loop Railroad. 187 – John Cornwell. 188, 189 – Ray Hooley. 190 – R. K. Evans. 192 – John Topham Picture Library. 194 – University of Reading, Institute of Agricultural History and Museum of English Rural Life. 195 – Guernsey Tomato Centre Ltd. – Road Locomotive Society. 196 – Beamish North of England Open Air Museum – University of Reading, Institute of Agricultural History and Museum of English Rural Life. 197 – Derek Petty. 199 – University of Reading, Institute of Agricultural History and Museum of English Rural Life.